GEOTHERMAL ENERGY IN THE USSR

A Survey of Resources, Methodology, Geology, and Use

Savely Polevoy

San Jose New York Lincoln Shanghai

Geothermal Energy in the USSR
A Survey of Resources, Methodology, Geology, and Use

iUniversity Press
an imprint of iUniverse.com, Inc.

For information address:
iUniverse.com
5220 S 16th, Ste. 200
Lincoln, NE 68512

Originally published by Delphic Associates Inc.

ISBN: 0-595-14938-3

Printed in the United States of America

ACKNOWLEDGEMENTS

The author wishes to express his gratitude to all who participated in the preparation of this monograph. Special thanks are due to Steven Jones for his extensive and painstaking editing. The author is further indebted to the technical advisor, Robert LaFleur, for his indispensable advice; finally, Masha Sinkevich is to be thanked for her splendid translation.

CONTENTS

ILLUSTRATIONS

iv

TABLES

FOREWORD

This monograph details research and development of geothermal energy resources in the USSR, describes geologic occurrence and technical state-of-the-art production methodology, and offers a prognosis for future growth in the Soviet geothermal industry. The author combines his geohydrological career experience with an extensive literature review to offer a comprehensive report that should interest geotechnical readers and also those who study the workings of the Soviet science and technology bureaucracy. Chapters 1 and 4 deal with where the geothermal industry has come from, the problems it faces, and where it is going. Chapters 2 and 3 detail the how and where of geothermal production, and include estimates of reserves.

State-wide support for geothermal research and development began in 1964 in the USSR. Prior to that time, more than 50 geothermal basins and volcanic sources had been identified, electric power generation potential had been suggested, and the first geothermal map of USSR had been prepared. A comparable nationwide effort of roughly the same magnitude was started at about the same time in the US.

In the late 1960s, low- and medium-potential geothermal waters, with temperatures less than 100°C, gained widest use. By 1972, 62 wells were in production and the Pauzhetka geothermal power plant was producing 15 million kW annually. This experimental plant was built to prove the feasibility of using steam-water mixture to generate electrical energy and its economic viability. Both were demonstrated. The hierarchy that was responsible for R&D support and growth during this early period, and still serves today, is described in chapter 1.

In 1979, thermal water energy provided heat for 4,500 apartments, hot water for 300,000 people, heated 50 hectares of hothouses, and generated 16 million kWh of electricity. In 1980, the Soviet Union was exploiting 36 geothermal water fields (GWFs) comprising 170 wells that produced 40 million m^3 of hot water. Steam production, which occurred only at the Pauzhetka power station, was 250-300 thousand tons annually in the period 1968-1980. Although they fell short of plan targets, these figures represent but a fraction of hypothetical Soviet thermal and superheated water reserves. Estimates of hypothetical thermal water reserves show the overall energy of these reserves to be on the order of 200 million Gcal annually.

All Soviet geothermal water fields (GWFs) are of the hydrothermal type where the liquid phase dominates. There is no production from hot dry rock (HDR) or magma sources because technology is insufficient. But then again, there exists no place in the world a technology for the production of energy from HDR. Two basic types of GWFs are exploited: stratal water-pressure systems, and fissure-vein systems. The stratal GWFs produce mainly from sedimentary formations that are widespread on platforms, in foredeeps, and in intermontane depressions where Mesozoic and Cenozoic deposits contain extensive volumes of thermal water, generally at 75-100°C. Fissure-vein fields produce from fracture systems that relate to on-going tectonic activity and volcanism where water temperatures may reach 300°C. Drilling technology in each field type is different. In stratal systems, wells are generally about 2,000 m in depth, and have bottom diameters averaging 152 mm. This depth is greater than the average depth of comparable US wells. In fissure-vein

systems, wells are in the 400-600 m depth range, and special precautions
are taken during their construction. Geophysical techniques used in
fissure-vein exploration include DC electro-profiling (bilateral dipolar), and correlative refractive wave profiling methods. In the US,
essentially similar methods and techniques are in use.

Stratal systems and formations are evaluated by hydrodynamic calculations and simulations based on wellhead data. These calculations
were the result of original research by the author, which yielded a unique method to determine small pressure differences in wells with a high
degree of accuracy. Flow tests on wells in the USSR are varied but are
similar to U.S. techniques; these may last several days or months, and
ideally lead to a 25-year forecast of well or field performance. Forms
of the Theis equation and elastic regime filtration theory are applied
to estimate performance of stratal wells; boundary problems and ani-
sotropy enter into calculations aimed at the 25-year forecast. These
analytical methods vary little from those used at cold water wells.
Simulations of well field performance include network electric (RC networks) and series analog methods.

Fissure-vein systems, however, are evaluated empirically. These
yield steam-water mixtures, are generally non-Darcian inflow behavior,
and their explorable reserves exceed natural discharge by a factor of 3
to more than 30. Well tests in fissure-veins systems usually last at
least a year to allow impact of the seasonal hydrologic cycle to be
seen.

Formulas for determining static and dynamic pressures from wellhead data, heat conductivity within the open borehole, and impact of
drawdown are presented. All of the evaluation methods presume a gushing
well production mode. Exploitation by pumping is currently economically

unfeasible in the USSR, but studies are continuing with the aim of developing an economical pump for geothermal wells. Chronic problems in geothermal wells include control of sand bridging, corrosion, and salt encrustation. Similar problems are encountered in other parts of the world (Hungary, Iceland, New Zealand, and elsewhere), where similar hydrogeothermic features are found.

Geothermal resources are widely spread across the Soviet Union in eleven geologic provinces. Structural, tectonic, and stratigraphic settings, well yields, and mineralization levels in each province are reviewed, and reserves are predicted.

In addition to the single operating power plant at Pauzhetka, other projects are underway in the Stavropol, Dagestan, and Transcaucasian areas. Current technology permits recovery of more than 20 kWh of electric energy per ton of water at 150°C.

Even so, geothermal electric plants are on the periphery of expected Soviet energy production. Because more than 90% of geothermal reserves are between 40-100°C, the most practical uses are for hot water supply for space heating, e.g., residential complexes and greenhouses, and for health spas--long a popular use in the Soviet Union.

Breakthroughs in hot dry rock (HDR) technology appear necessary if there is to be significant increase in geothermal production. The development of high temperature underground boilers in the USSR will be realized only if a number of hydrological, seismic, and economic problems that relate to the drilling and casing of superdeep boreholes in high temperature conditions (300-500°C) can be solved. The USSR has developed an installation for drilling superdeep boreholes to 15 km: the Uralmash 15000. However, the extremely small bottomhole diameter it produces is inadequate for creating underground boilers at great depths.

Nonetheless, the 1990s are expected to see production of conductive HDR heat in volcanic regions of Kamchatka and permafrost areas of the northeastern USSR via underground thermal boilers lying at depths of up to 3 km. By the turn of the century, high-temperature (400-600°C) thermal boilers in HDR at depths of 6-15 km should come into use.

In the USSR no separate administrative structure for geothermal energy has been developed. Exploratory efforts for thermal water are still carried out by two ministries: the Ministry of Geology and the Ministry of the Gas Industry, while extraction and sale of thermal water is conducted solely by the latter. The lack of an independent organization hinders geothermal development and results in irregularities in planning, distribution of resources, and allocation of materials and equipment.

Robert G. LaFleur
Department of Geology
Rensselaer Polytechnic Institute

INTRODUCTION

The ever-increasing consumption of non-renewable fuels of organic origin--coal, oil, gas, peat, and oil shale--hastens the exhaustion of these resources. Available data indicate that if the extraction of organic fuels continues at the same pace as at present, the reserves of these resources will be significantly depleted in approximately 80 to 100 years. Consequently, the interest in non-traditional renewable resources such as the sun, tides, wind, and geothermal energy is understandable. The inner heat of the Earth, which is believed to result from the radioactive decay of elements, theoretically represents a virtually inexhaustible energy resource. It has been estimated that the total heat reserves contained in the Earth's crust to a depth of 10 km comprise some 10^{27} joules, a figure which exceeds the calorific value of world coal reserves by a factor of 2,000 [7].

Despite the impressive quantity of geothermal reserves, large-scale commercial exploitation is hampered by a series of obstacles. One of the greatest of these impediments is the form of the resources: natural steam formations, for example, which are the most valuable from of geothermal heat, are exceedingly rare. The overwhelming majority of the Earth's geothermal reserves are found in two types of formations: geothermal water fields (GWFs) and hot dry rock (HDR); principal reserves are concentrated in the latter form as conductive heat. At present, the extraction of heat from HDR is not commercially viable due to a number of unresolved difficulties. As a result of these and other factors, geothermal R&D efforts worldwide have long been focused on the use of GWFs.

Although they are considerably less extensive than are HDR resources, GWF energy resources are not insignificant: in the last two decades, geological research has estimated the hypothetical thermal water reserves of the USSR alone to offer some 200 million Gcal annually for an indefinite period of time [16]. If fully exploited, these resources would represent considerable savings in fossil fuels. Annual savings for the Soviet Union would be on the order of 40-50 million tons of oil [10].* This is enormously attractive because oil is the principal hard currency export of the USSR. Moreover, benefits obtainable from GWFs are by no means restricted to the field of electrical energy production alone. The use of geothermal waters for balneologic purposes has been known since ancient times, and in this century numerous other applications have been revealed, such as heating for buildings; in agriculture, geothermal water has been successfully used for heating greenhouses and hothouses.

The exploitation of geothermal water is not without difficulties, however. Unlike, for example, electricity, geothermal energy is not transportable over great distances: unless it is converted to another form of energy, geothermal energy must be consumed locally. Although thermal waters are broadly distributed over the territory of the USSR, the most significant resources are found far from the European part of the country. Furthermore, all fields are not equal from the standpoint of commercial use. The feasibility of exploiting geothermal heat is determined by natural hydrogeothermal and geologic factors that necessitate comprehensive investigation of the distribution and quality of

---

*Oil and gas-condensate production were expected to reach 628 million tons in 1984 (Pravda, 28 November 1984).

2

geothermal waters. The usefulness of exploitable thermal resources is stipulated primarily by their dimensions and energy potential; this potential is related to the formation of natural water-pressure systems in highly heated rock.

Large-scale use of exploitable geothermal water resources is contingent upon overcoming technical problems. The development of geothermal energy resources has also brought about the need to resolve numerous issues related to the discovery of thermal water fields that are found at great depths and are not evident from the surface. Additional obstacles are mineral deposition and corrosion, finding the most effective means for heat collection, and developing inexpensive and highly productive drilling techniques for drilling geothermal wells.

Such difficulties notwithstanding, the Soviet geothermal effort has not been without success. The first step in the development of commercial use of geothermal heat in the USSR was initiated in the late 1950s with the construction of a geothermal electric power plant on Kamchatka: this plant went on-stream in 1967. Nonetheless, 1964 should be viewed as the start of large-scale Soviet research on geothermal energy use. Geologic prospecting operations were initiated in many regions of the country to develop already known sites and identify new geothermal fields. The sixties were a period when the use of thermal waters for heating in cities, for agriculture, as well as to produce electric power, represented a new trend in Soviet energy and industry. In the last twenty years, the Soviets have carried out a broad range of practical scientific research; this research facilitated the comparatively rapid development of a number of geothermal fields and the identification of new fields with geothermal energy use potential. This trend in

Soviet energy continues: in 1979, the USSR produced 40 million m^3 of thermal water and 300,000 tons of steam, with enthalpy equal to some 1 billion kWh.

This monograph seeks to shed light on all aspects of Soviet geothermal energy developments related to the exploitation of GWFs--hydrogeothermal, technological, economic, as well as research methodology and evaluation of geothermal sources. Chapter I presents an overview of the research that resulted in the establishment of a new fuel industry sector in the Soviet Union. The chapter also outlines the organizations performing scientific, research, geologic, and prospecting operations, as well as the development of geothermal waters and their extraction. Chapter II examines standard Soviet geologic and prospecting techniques and issues concerning the search for subsurface thermal waters, exploration of hydrogeothermal fields, and the evaluation of their resources. Chapter III provides a hydrogeothermal characterization of Soviet thermal water basins with commercial potential, as well as that of explored geothermal fields. Also included is the country-wide distribution of the USSR's thermal resources. Chapter IV is dedicated to the current and future use of geothermal energy sources: this chapter examines such issues as use trends, technological problems tied to exploitation of geothermal sites, and the economic effectiveness of geothermal systems. The final chapter further presents some recent ideas on use of the geothermal energy and a brief assessment of prospects for the future of the USSR's efforts to exploit this renewable resource.

CHAPTER I

OVERVIEW OF THE ORGANIZATION AND ADMINISTRATION OF SOVIET GEOTHERMAL
RESEARCH AND DEVELOPMENT

Developments in Soviet research on geothermal heat use have been
intimately connected to advances not only in exploration efforts and
technique, but also to the accumulating body of data they have facilitated. For this reason, the evolution of the R&D structure may be best
understood when viewed in the context of these achievements. This
chapter begins with a description of the general course of these
achievements, including the more noteworthy milestones and trends.
Thereafter follows a view of the organizational mechanism which has
resulted as the culmination of some twenty years' effort to expand the
domestic geothermal program. The chapter is then concluded with a sum-
mary briefly outlining the manner in which this apparatus typically
functions.

Planned systematic geothermal research in the USSR commenced in the
mid-1950s. Of course, temperature observation of data had been gradually accumulating since much earlier, beginning with the first soil
temperature measurements taken in the 1730s, and later expanding to
include data for thermal sources, mines, and wells. The first theoretical and experimental studies on the Earth's thermal field date from this
early period as well. These efforts, however, were confined solely to
sporadic exploration of geothermal resources and phenomena, that is,
exploitation had not yet received concerted attention.

It was at the First All-Union Conference on Geothermal Research, held in Moscow in 1956, that the commercial use of geothermal resources first came under discussion. From this point forward systematic research on geothermal energy has been conducted in the USSR. In the wake of the Conference came a deluge of varied research activity. The first references to thermal water as a heat source found their way into print, as did Kraskovskiy's generalized works on geothermal gradient measurements in the USSR; Kraskovskiy's data made it possible to plot regional geothermal maps and cross sections. It was during this period that the first schematic geothermal maps were prepared.

By 1964--which was to be the next milestone in the Soviet geothermal program--significant generalized works treating the distribution and formation of the subsurface geothermal field and thermal waters within the territory of the USSR were available. Furthermore, the development of deep oil and gas exploratory drilling, in tandem with the generalization of geothermal drilling measurement data, demonstrated the presence of substantial geothermal energy reserves in the USSR. This resource was manifested in water overflows of varying temperature, as well as hydrothermal steam springs. More than 50 thermal water basins capable of providing heated water were identified, and the major potential for developing the heat resources of volcanic regions was discovered, including their use for electric power production.

The results of theoretical and practical investigations conducted during the eight years since the first Conference were summarized at the Second All-Union Conference on Geothermal Research, which was held in Moscow in March of 1964. The greatest advances had been achieved in mapping underground temperature characteristics and identifying hydrothermal resources suitable for heat and power generation. The first consolidated geothermal map of the USSR (1:5,000,000) had been compiled through combined efforts of several institutions: the Union Republic Academies of Sciences, the Geologic Institute of the USSR Academy of Sciences, and the Ministry of Geology and its territorial administrations. A preliminary evaluation of commercial-level hydrothermal resources had been carried out, and territories with hydrothermal water potential had been identified on Kamchatka and in the Caucasus, Central Asia, Kazakhstan, Western Siberia, and other regions.

Mapping was not the only facet of geothermal energy to receive attention from 1956 to 1964, however. Technical aspects of geothermal heat use were addressed, such as geothermal electric power plant construction and plans for the use of thermal water for space heating and in agriculture; general theoretical problems of the Earth's thermal features and balance in relation to the planet's thermal evolution were examined as well. Moreover, research on the interrelationship of the Earth's thermal field with other geophysical fields and geothermal processes was conducted.

Despite the obvious successes of this first period, the greatest boost to the Soviet geothermal program was yet to come. The stimulus was provided by a 1963 Council of Ministers Decree*; as a result of this high-level policy decision—which added state-wide support to the program—1964 was to mark the beginning of a new period in the Soviet geothermal program. The geothermal effort was expanded both in scope and intensity, and for the first time it was given a decidedly practical

---

*"The Development of Operations for the Utilization of Geothermal Heat in the National Economy" was signed by A. N. Kosygin, then Deputy Chairman of the USSR Council of Ministers.

slant: this year heralds the start of systematic and intensive geologic, exploratory, and construction operations for development of existing sites and prospecting for new sources of geothermal energy.

In connection with the new impetus given to the Soviet geothermal program, in 1964 the USSR Ministry of the Gas Industry (Mingazprom) and the Ministry of Geology jointly undertook a coordinated plan for drilling prospecting and exploratory geothermal wells in Kamchatka, the Caucasus, Central Asia, and Kazakhstan. By 1971, 0.4 million m of prospecting and exploratory drilling for thermal waters had been carried out under the auspices of this program [8]. Also in 1964 Mingazprom commenced operations to recover wells at abandoned oil fields in Dagestan and the Chechen-Ingush ASSR, with the purpose of connecting these wells with promising thermal water beds. The efforts to reconstruct abandoned oil wells did not require significant investments of capital or time, which made it possible to prepare an industrial evaluation and begin exploitation of thermal waters in a comparatively short period (1964-66) in the Makhachkala, Ternair, and Izberbash areas (Dagestan ASSR) and the Oktyabr'skaya area (Chechen-Ingush ASSR).

Simultaneously, a need arose to improve methods for guiding search and exploratory efforts and evaluating the presence of the reserves with commercial potential. For the first time, economic evaluations of the use of Soviet thermal resources were conducted; these evaluations were made for regions with varying economic structures and different climatic conditions. Included among these studies were assessments of the economic effectiveness of exploiting thermal water and steam-water mixtures of varying temperatures; the results from these studies could be compared with data for other heat sources. The

8

research and feasibility studies performed in the 1960s were to be used in establishing home heating systems and providing hot water to cities, work settlements, industrial concerns, and for farming, utilizing thermal waters of Kamchatka, Dagestan ASSR, Chechen-Ingush ASSR, Transcaucasus, South Kazakhstan, and West Siberia.

Since low- and medium-potential thermal water (temperature to 100°C) has gained the broadest use in the Soviet Union, the focal point of practical development of thermal water resources lies in the most effective use of heat carriers with these thermal parameters. In order to study the possibility of converting low-potential thermal water heat energy to electrical energy, the Leningrad Technological Institute for the Refrigeration Industry, jointly with the Thermophysics Institute of the Siberian Division of the USSR Academy of Sciences, designed a freon turbo-unit with a closed low-boiling heat carrier loop (Freon-12). The first experimental turbine of this type with a 450 kW capacity went on line at the Paratunka thermal water field on Kamchatka in September, 1967. Water temperature at wellhead did not exceed 81.5°C.

Another effort was the development of efficient heating installations and electro-turbines, as well as thermal insulation systems and anti-corrosion devices. There were studies concerning the forecasting of changes of pressure, flow rate, temperature, and chemical composition of thermal waters during exploitation of geothermal basins. Research was conducted on how to protect geothermal wells and surface thermal pipelines from mineral deposition, and specifications for chemical composition, overall mineralization, level of warmth, and quantity of thermal waters were developed in order to evaluate the feasibility of using these sources for energy production and other uses.

Of the many projects set into motion in the 1960s were also several focusing on the use of superheated water for the production of electric power. One of these, undertaken near Makhachkala in the Dagestan ASSR, was the drilling of an exploratory geothermal well 4,000 m deep; plans envisioned obtaining a steam-water gusher from this well with a wellhead temperature of 140°C. Another milestone project was the detailed exploration and industrial evaluation of the Bol'she-Bannoye superheated water field on Kamchatka; this area was to be the site for the construction of a 25,000 kW geothermal power plant.

During the 1960s, measures were also developed for commercial production of vegetables by the Ministry of the Gas Industry jointly with the Scientific Research Institute for Vegetable Growing of the USSR Ministry of Agriculture. These measures included construction of test-commercial hothouse and greenhouse combines on Kamchatka, in the Stavropol area; in the Georgian, Kazakh, and Tajik SSRs, and in the Dagestan and Chechen-Ingush ASSRs. Thus, the following targets were planned for the central region of Paratunka Springs in Kamchatka: 9,000 m^2 of winter greenhouse area, as well as 4,600 m^2 of summer greenhouses; 6,670 m^2 of hothouses; 21,000 m^2 of covered heated ground, and 17,000 m^2 uncovered. Substantial areas of greenhouses and hothouses were planned in the area of Pauzhetka, the Caucasus and in Central Asia (altogether, more than 100,000 m^2).

As a result of such practical efforts to develop geothermal resources, the foundation for using thermal waters in the Soviet Union was laid by 1966. Exploitation of the first thermal water facilities was initiated in the Caucasus (in the cities of Makhachkala, Izberbash, Groznyy, Tbilisi, Zugdidi), and on Kamchatka (near the city of Petropavlovsk). In 1967, the first geothermal electric power plant in the USSR, the Pauzhetka GeoTES (situated on the superheated water field of the same name), was put on stream. The initial capacity of this plant was 5000 kW.

Toward the end of the 1960s, the extraction of thermal waters was substantially increased, especially in the Caucasus, as a result of recovery of abandoned oil wells and reconstruction on their thermal water strata at already developed sites, such as Makhachkala, Ternair, and Groznyy. The increase was also due to development of newly explored GWFs, such as Kizlyar (Dagestan ASSR), Cherkessk (Stavropol area), and Okhurey (Abkhaz ASSR). In 1971, Soviet geothermal well stock consisted of 62 wells, and in 5 years, production of thermal waters had increased from 0.375 million m^3 in 1966 to 8.4 million m^3 in 1971. Steam production reached 250,000 tons in 1971. In the same year, 50,000 non-industrial consumers were using thermal waters; thermal energy heated 35,000 m^2 of residential and manufacturing area, 150,000 m^2 of greenhouses and hothouses, including the Paratunka Greenhouse-Hothouse Combine (with an area of 60,000 m^2). At this same time, electric power production at the Pauzhetka Geothermal Power Plant comprised 15 million kW annually. Explored industrial reserves of thermal waters and water-steam mixture were 110,000 m^3/day and 14,600 tons/day, respectively. In the five-year period 1966-1971, Soviet industrial exploitation of thermal waters replaced 70,000 tons of conventional fuel, including 50,000 tons for 1970 and 1971 [8].

Meanwhile, the deep geothermal well at the Karaman area (Dagestan ASSR) mentioned above, which was drilled to a depth of more than 4,000 m, resulted in an insignificant overflow of highly mineralized water with a wellhead temperature of approximately 50°C. Consequently, the well was

11

abandoned. Exploratory wells drilled in the 1960s south of the city of Makhachkala along the shoreline of Dagestan also failed to produce industrial-level flow rate. Further efforts toward exploiting the Bol'she-Bannoye superheated water formation (with exploitable steam-water mixture reserves of 250 kg/sec, temperature to 170°C) in Kamchatka were suspended. According to a feasibility study compiled in 1967, construction of the Bol'she-Bannoye GeoTES with a 25 MW capacity was inadvisable due to the high capital investments required for its construction, as well as due to the small share of necessary electric power that it would cover in the Petropavlovsk power district.

Plans were to bring production of thermal waters to 45 to 50 million m^3 annually by 1980, and that of steam and water to 2 million tons [8]. Also anticipated was the exploration and bringing on stream of new thermal water fields on the territory of Kamchatka, the Kuriles, Sakhalin, Kazakhstan, Central Asia, and Azerbaijan. By 1980, annual savings of conventional fuel were expected to reach 170,000 tons: the use of geothermal heat in the Soviet Union had acquired a production base and become a noticeable supplementary source of energy for the country.

The great strides made in the Soviet geothermal program in the 1950s, 1960s, and 1970s would not have been possible without the support of numerous scientific technical institutions. The development of the organizational and administrative framework for geothermal R&D in the USSR underwent great development parallel to that of the theoretical and applied research outlined above.

*   *   *

In the Soviet period, prior to the first All-Union Conference on Geothermal Research in 1956, many Soviet scientific and production organizations conducted research on geothermic issues related to oil and gas production, prospecting and recovery operations, and the study of the geothermal mechanisms of the subsurface cryosphere. Of significance are works by A. N. Ogil'vi (1932) concerning the role of geothermy in hydrogeologic research, and V. G. Khlopin and A. N. Tikhonov's work (1937) on the effects of radioactive media on the thermal characteristics of the Earth. Principal theoretical research on geothermal heat and geothermic mechanisms in the depths were developed in the Earth Physics Institute of the USSR Academy of Sciences (S. A. Kraskovskiy, I. D. Dergunov, Ye. A. Lyubimova). Research was conducted at the Moscow Petroleum Institute on geothermal studies of exploratory oil wells (V. N. Dakhnov, D. I. D'yakonov), as well as the geothermy of oil and water-bearing strata as related to secondary methods of oilfield production (I. A. Charnyy). The study of Kamchatka's geothermal resources was carried out by the Volcanology Laboratory USSR Academy of Sciences (B. I. Piip, S. I. Naboko, T. I. Ustinova). Research on the characteristics of numerous geothermal mineral (medicinal) sources was conducted by associates of the Health Resort Treatment Institute of the USSR Ministry of Health (V. V. Ivanov, et al.) and its local branches. During this period, research on geothermal energy use was not goal-oriented; research was conducted separately by individual researchers and lacked overall coordination.

In September 1961, the USSR Academy of Sciences Presidium formed a Commission for Hydrogeology and Geothermy to coordinate efforts in promising scientific and production directions. In order to achieve syste-

13

matic development and coordination of geothermal research, the USSR
Academy of Sciences initiated the second All-Union Conference on
Geothermal Research (March, 1964), in which representatives of 60 scientific and production organizations participated.

After the Conference, scientific institutions had their existing
programs expanded to include geothermal research. Study of geothermal
applications and technology was initiated during this period at the
Thermophysics Institute of the Siberian Division of the USSR Academy of
Sciences (S. S. Kutateladze, V. N. Moskvicheva); later this institute
developed the first Soviet freon turbine (1967). Study of thermal
waters as an energy source was initiated at a number of institutes: the
USSR Academy of Sciences Geology Institute (F. A. Makarenko, B. G. Polyak, A. B. Shcherbakov), the USSR Academy of Sciences Geology Institute
for Ore Deposits, Petrography, Mineralogy and Geochemistry (I. M.
Dvorov), the Groznyy Petroleum Institute (G. M. Sukharev), the USSR
Ministry for Energy and Electrification, and the Kamchatka Geologic
Administration (Yu. A. Krayevoy). Research was expanded by the USSR
Academy of Sciences Volcanology Institute (V. V. Aver'yev, V. M. Sugrobov, A. A. Gavronskiy). In order to coordinate the expanded research
and geologic exploratory efforts after 1964, the Scientific Council on
Geothermic Research was established under the Earth Sciences Division at
the USSR Academy of Sciences.

Since geothermal energy exploitation was viewed as a new branch
of the fuel industry, the execution of principal geologic exploratory
and construction operations was assigned to the USSR Ministry of the Gas
Industry (headed by V. Kortunov). In the spring of 1964, the Division
for the Utilization of Geothermal Energy (Division Chief V. A. Nursha-

nov, Chief Geologist Ye. F. Bolgarina) was established under the
Ministry. This division included drilling, hydrogeology, and heat technology specialists; its responsibility included the overall management
and coordination of research and exploratory geologic operations related
to the development of geothermal energy performed under the auspices of
Mingazprom. The operations themselves were conducted by the All-Union
Natural Gas Science and Research Institute (VNIIgaz) (L. M. Zor'kin),
the North Caucasian branch of VNIIgaz (I. P. Kovshov and the author),
and the Vostok Giprogaz State Planning and Research Institute (A. A.
Shpak). Also in 1964, the Mingazprom Caucasian Exploratory Expedition
was set up at Makhachkala, and its responsibilities included the
drilling of geothermal wells in the area of the North Caucasus and
Dagestan.

A substantial volume of geothermal well drilling operations in the
Far East (Kamchatka and the Kurile Islands), in Central Asia, and in the
Caucasus was carried out by the USSR Ministry of Geology. This minis-
try's scientific and research efforts for the study and evaluation of
thermal water resources were conducted by the following organizations:
the All-Union Science and Research Institute for Hydrogeology and
Engineering Geology (VSEGINGEO) (V. F. Mavritskiy, N. M. Frolov, M. A.
Ogil'vi, and J. K. Antonenko); the Uzbek SSR Mingeo Hydrogeologic Trust
(G. V. Kulikov, and T. B. Grebenshchikova), and the Central Asian
Scientific Research Institute for Geology and Mineral Ore (B. A. Beder).
In the 1960s, the following academic institutes joined in the effort to
study the utilization of thermal water:

> -the Dagestan Scientific-Research Division of Energy
> (S. A. Djamalov);

> -the Geology Institute of the Dagestan Branch of
> the USSR Academy of Sciences (M. K. Kurbanov, S. A.
> Kasparov);

-the Azerbaijan SSR Academy of Sciences Geology
Institute (M. A. Kashkoy, S. A. Aliyev);

-the Kazakh SSR Academy of Sciences Institute for
Hydrogeology and Hydrophysics (U. M. Akhmedsafin,
V. S. Zhevago);

-the Lithosphere Institute, Siberian Division USSR
Academy of Sciences (I. S. Lomonosov, S. V. Lysak);

-the Inorganic Chemistry Institute, Siberian
Division USSR Academy of Sciences (P. A. Kryukov,
E. G. Larionov);

-the Geology Institute of the Ukrainian SSR Academy
of Sciences (V. I. Lyal'ko, A. Ye. Babinets, M. M.
Mitnik);

-the Technical Thermophysics Institute of UkrSSR
Academy of Sciences (O. A. Kremnev, A. V. Shurch-
kov);

-the Laboratory for Geochemical Problems of the
Belorussian SSR Academy of Sciences (G. V. Bogo-
molov).

Research efforts to study thermal water were also carried out by

some research centers in institutions of higher learning, such as the

Georgian Polytechnical Institute (I. M. Buachidze, M. P. Shaorshidze);

the Groznyy Petroleum Institute (S. P. Vlasova, Yu. K. Taranukha); Azer-

baijan State University (A. G. Askerov); and Leningrad's Plekhanov

Mining Institute (Yu. D. Dyad'kin, Yu. M. Pariyskiy, A. V. Vaynblat).

Feasibility studies for geothermal heating systems were carried out

by USSR Gosstroy Central Scientific Research Institute for Experimental

Equipment Design (TsNIIEP) (B. A. Lokshin, A. V. Vol'fenfel'd); and the

USSR Ministry of Energy (Minenergo) Teploproyekt Institute (G. A. Myslin,

Yu. Ya. Geller). Issues concerning utilization thermal waters for power

production were studied at the Krzhizhanovskiy Energy Institute (B. I.

Kozlov), and research on the use of thermal water was conducted by the RSFSR Ministry of Agriculture Scientific Research Institute for Vegetable Farming.

In 1966, the Caucasian and Kamchatka Industrial Administrations for Utilizing Geothermal Heat were established by Mingazprom in Makhachkala and Petropavlovsk-Kamchatskiy, respectively. They were charged with organizing the production of thermal water at identified hydrothermal formations in their areas. Later, in 1968, the Caucasian Industrial Administration for Utilizing Geothermal Heat was separated into several divisions: (1) the Caucasian Division, which outfitted geothermal fields, carried out extraction and supplied thermal water to consumers within the Dagestan and Chechen-Ingush ASSRs; (2) the Georgian Division (in Tbilisi), which is responsible for the same operations on Georgian territory; and (3) and the Northern Caucasian Division (in Mineralnyye Vody, near Stavropol), which exploited thermal waters in the Kabardino-Balkar ASSR and in the Stavropol and Krasnodar administrative regions. Figure 1 shows the administrative hierarchy of the key organizations currently involved in the administration of Soviet geothermal R&D.

Exploration of geothermal reserves falls under the adminstrative authority of Mingazprom and the Ministry of Geology: the direction and the extent of the Ministries' geothermal efforts are thus influenced by the directives of the Council of Ministers and the budget allocated by Gosplan. An initiative concerning exploration and exploitation of geo-thermal resources in a given area originates on the ministerial level. The directive for an investigation is transmitted to geologists in the local administration in question, who conduct the study and present their results in the form of a geologic report (geologicheskiy otchet).

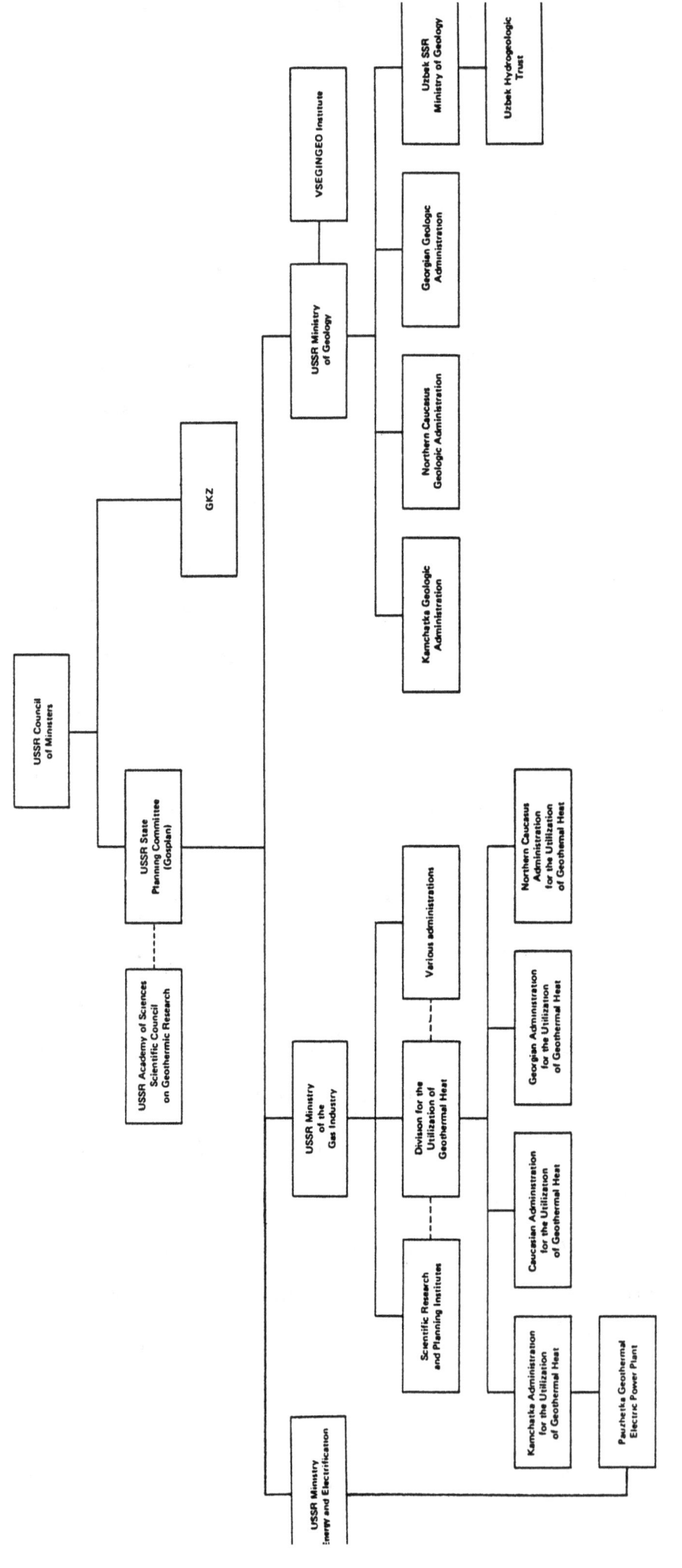

FIGURE 1

ORGANIZATIONAL STRUCTURE OF SOVIET GEOTHERMAL ENERGY RESEARCH AND USE

The geologic report includes an assessment of thermal water reserves, but does not treat the issue of economic feasibility, which is addressed later. Such reports are first submitted to the State Commission on Mineral Resources (GKZ), which is subordinate to the Council of Minsters. The GKZ, which examines and approves reports on mineral reserves, has a permanent staff augmented by consultants drawn from academic institutions. Its costs are covered through the budgets of the given geologic administrations. After approbation by GKZ, the report is returned to Mingazprom with a commercial evaluation (protokol) of the site describing both the reserves and suitable applications for them.

If the Mingazprom decides to pursue the project, it contacts the appropriate design organization; for example, in the case of an electric power plant, the Ministry would arrange for the Ministry of Energy to conduct an economic feasibility study. At this stage it is determined if the resources can be exploited economically and exactly what form the exploitation should assume. These results are returned to Mingazprom, which then arranges the construction operations through the appropriate organizations.

As a result of geohydraulic and other research, by the end of the 1960s the commercial effectiveness of thermal waters in the USSR had been demonstrated in the Caucasus, Kamchatka, Central Asia, and Siberia. Four industrial adminstrations had been established, and research on geothermal energy use had taken on a planned and long-term character.

CHAPTER II

GEOTHERMAL METHODOLOGY

Issues related to the origins and behavior of geothermal heat are
extremely complex and still remain inadequately researched. Meanwhile,
the hypothesis of the steady state of the Earth's inner heat flow for
the historical period and its radiogenic origins is supported by a
majority of scientists. The methods that are used to explore and evaluate geothermal water fields* (GWFs) in the Soviet Union are based on
this understanding of the origins of the geothermal heat and the formation of the geothermal reservoirs.

Exploratory procedures to be conducted at a geothermal site are
selected in accordance with the following factors: the geologic structure, the local and regional hydrodynamics of the water pressure system
to which the GWF belongs, the degree of heat, and the water phase state
in the strata and movement along the well stem. This chapter provides a
classification of GWFs found in the USSR, a description of the sequence
of Soviet geologic and exploratory operations, and the classification
system used for geothermal reserves in the USSR. Thereafter follows a
description of drilling operations used in the exploration of geothermal
fields and areas, and finally, a discussion of Soviet geophysical and

_____

*The term "geothermal water field" as defined in Soviet hydrogeological
literature means a spatially limited accumulation of underground water
whose temperature, chemical composition, and exploitable reserves
satisfy requirements for production of power or heat, and whose
exploitation is feasible with the use of contemporary equipment and
technology.

hydrogeothermal research related to prospecting, exploration, and evaluation of GWF resources.

1. Classification of Geothermal Formations

In order to explain how Soviet geothermal areas are studied, the author uses a general classification method provided by Edwards [11], which classifies geothermal fields by reservoir type:

(1) hydrothermal (liquid-dominated and
    vapor-dominated)
(2) geopressurized
(3) hot dry rock (HDR)
(4) magma

Liquid-dominated hydrothermal reservoirs are hydrothermal reservoirs that contain circulating heated water in a droplet-liquid state; this water is the principal carrier of geothermal heat. Water temperature in such systems may significantly exceed the boiling point of water under normal conditions (100°C at 1 atm pressure), i.e. the water may be in a superheated state in the presence of hydrostatic pressure which significantly exceeds atmospheric pressure. All explored and presently exploited Soviet GWFs are of this type. Even in the Pauzhetka formation, whose wells produce water and steam (formation temperatures to 200°C), the water in the strata is in liquid phase. Steam formation occurs in the well stem when hydrostatic pressure is reduced [12].

Vapor-dominated (steam) hydrothermal reservoirs are dry steam formations. Such dry steam formations are extremely rare and are known only in Italy (Larderello, Monte Amiata), in the US (California), and in Japan (Matsukawa). Temperatures in these formations are in the 250-300°C range.

Hot dry rock (HDR) is known to be found widely as a result of accumulated drilling experience that has established that temperature increases with depth. At a depth of 1,000 m below the Earth's surface, strata heated to 30°C are usually found, while in recent fold systems and in regions of contemporary volcanic activity the highly heated strata may lie at a depth of several hundred meters. In Epipaleozoic platforms, the highly heated strata are found at a depth of several thousand meters. HDR reservoirs occur in heated strata where the rock is impermeable or slightly permeable by water; these reservoirs are significantly more common than reservoirs containing thermal water. The thermal energy locked in the Earth's crust (to a depth of 10 km and at temperature exceeding that of the surface) amounts to $3 \cdot 10^{26}$ cal [13]; at the present time, however, there are no effective direct methods for collecting the heat of such reservoirs. Extraction of thermal energy from HDR may be developed by the creation of artificial circulation systems.

Geopressured reservoirs are usually found deep in sedimentary basins. These are generally isolated sections of water-pressure systems in which the rock is heavily compacted and formation pressure substantially exceeds hydrostatic pressure. At such water-pressure sites, the geothermic gradient is usually above normal because of the increased specific heat capacity of super-compacted rock. In one such system on the shore of the Gulf of Mexico, a record-high temperature of 237°C was measured. Zones with abnormally high formation pressures have also been discovered in the USSR: these pressures exceed normal levels by 200 to 300 atm (Galyugayevskaya Well in the Chechen-Ingush ASSR, Praskoveyskaya Well near Stavropol, and others). Thermal water temperatures in these areas is 80-90°C. Knowledge currently available on the geologic structure and hydrodynamic features of areas with anomalously high hydrostatic pressures in the USSR shows that such sites do not usually contain large exploitable thermal water reserves.

Magma intrusions found at small depths in areas of volcanic activity are also natural reservoirs of potentially high thermal energy. However, the technology to utilize this energy has yet to be developed.

The most important GWF parameter that determines thermal potential and use of the reserves is the warmth of thermal water. Thermal water temperature fluctuates over a wide range; depending on this fluctuation, the temperature, and the formation pressure, thermal water is characterized by a different phase-aggregate state, composition, and properties. Therefore, a temperature criterion is used to sub-categorize thermal water based on its qualitative features in many classification systems.

The boundary between warm and hot water is 37°C (the temperature of the human body is 36.6°C), and that between hot and very hot water is 50°C (as the intermediate level in water used for medicinal and thermal energy purposes). Fumaroles, hot springs, and steam jets in excess of 100°C are placed in a separate groups termed "vapor-hydrothermae" and "steam-water mixture."

Superheated fluid signifies the state when temperature exceeds critical levels (in excess of 375°C). In a supercritical state, the aqueous fluid is in the form of highly dense gas; the density approximates that of liquid water. For practical purposes, the following classification of thermal water is used in the USSR:

Potential	Temperature, °C
Low	40 - 70
Average	70 - 100
High	> 100*

---

*Temperatures greater than 150°C are required for electric power generation.

Thermal water is also classified according to the level of mineralization. This classification illustrates in an arbitrary fashion the problem of waste water after collection of heat, as well as the corrosion properties and potential mineral deposition during exploitation of geothermal wells. The following classification for thermal water, based on the level of mineralization (g/l), is universal:

Fresh water	to 1
Brackish water	1 to 3
Salt water	3 to 35
Brines	over 35

Thermal water is also classified according to its ionic composition, which determines the water's chemical type in accordance with the predominant ions. Soviet GWFs are mainly classified according to geologic setting, structure, and hydrodynamic conditions.

In evaluating exploitable thermal water resources, the features of the geology and structure of GWFs and the nature of circulation of the thermal waters are particularly important. Two basic types of water-pressure systems and types of commercial GWFs related to them have been established on the basis of the following criteria [16]:

1. Geothermal water fields are limited to water-pressure systems of depression structures contains varying sedimentary formations. These are characterized by availability of stratal-pore and stratal-fissure thermal water.

2. Geothermal water fields are limited to folding regions of varying age that have undergone intensive recent tectonic and volcanic activity. Thermal water in these fields has small areal extent, occur in fractures, and is of the fissure-vein type.

On the territory of the USSR, water-pressure stratal systems are widespread in platforms, foredeeps and intermontane depressions where Paleozoic, Mesozoic, and Cenozoic deposits contain extensive areas of thermal water. In such systems, the maximum formation temperature is

usually 75-100°C, and rarely exceeds 200°C. In fissure-vein water-
pressure systems that are tied to fold structures of differing age,
thermal water circulates along a complex system of tectonic fissures.
Thermal water of this type found in areas of contemporary volcanic activity may be heated to a temperature of 300°C.

Geologic and exploratory efforts for thermal water include the most
diverse methods (e.g., geophysical, geological and hydrogeological survey methods, observation of the characteristics of thermal water
springs, etc.). However, the principal source for obtaining reliable
information for calculating a GWF's exploitable resources are drilled
wells and the tests carried out in them. Therefore, efforts directed
toward exploration and evaluation of a formation are carried out in a
staged, sequential manner. An evaluation of the prospects for locating
a GWF on a given territory is formed on the basis of all available data
on regional geological and hydrogeothermic conditions. Further efforts
seek to establish the type of GWF, the stratal conditions, and the gene-
sis of the resources. After data from earlier exploratory procedures
are analyzed, a feasibility study (tekhniko-ekonomicheskoye obosnovaniye)
is prepared for the region where the exploratory operations will be
carried out. This serves as the foundation for setting up and planning
prospecting and exploratory efforts. There are two basic stages of
geologic-exploratory exploration efforts (including GWFs): (1) exploratory search, and (2) prospecting (test drilling, and defining the commercial reserves of the GWF).

The basis for carrying out all stages of geologic and exploratory
efforts at GWFs is estimation of their reserves. Reserves of subsurface water (including thermal) are subdivided into natural (undisturbed)

25

and exploitable reserves. Natural reserves are divided into three sub-
types:

(1) Static reserves are the volume of subsurface water
    contained in water-bearing strata in the natural
    state, i.e., the overall volume of subterranean
    water found in pores and fissures in water-bearing
    rock. Static reserves may be extracted from the
    stratum only when it is drained.

(2) Elastic reserves may be extracted from the stratum
    by reducing hydrostatic pressure which causes
    deformation of both the water (when pressure is
    reduced water expands) and the water-bearing rock
    (rock becomes more dense and the water is squeezed
    off).

(3) Dynamic reserves (resources) are the waters in
    natural circulation. Dynamic reserves are usually
    exploited simultaneously with elastic reserves when
    hydrostatic pressure in the reservoirs drops.

In Soviet usage, exploitable reserves of subsurface (including
thermal) water consist of the yield of subsurface water that can be
extracted from wells and other installations in a technically and economically rational fashion, under the given conditions of exploitation
that maintain optimal quantity and quality. Such exploitable reserves
may not be limited to elastic and dynamic reserves in the exploited
water-bearing horizons. During extensive withdrawal of subsurface
water, changes occur in the water budget of the water-bearing unit. Due
to the fact that a cone of depression forms during the exploitation of a
GWF, conditions are created for drawing additional water resources to
the exploited area that were not taken into account when the natural
conditions were evaluated.

Examples of genetic characteristics of the additional reserves of
thermal water that can form in a GWF include leakage from neighboring
water-bearing strata, and concentrated supplementary feeding through

high-permeability rock that allows improved recharge (so-called "lithologic windows"). Additional reserves can also be formed in a fissure-vein GWF by drawing on surface water flows.

The exploitable reserves of a GWF according to the Soviet interpretation are typically represented by the following equation:

$$Q_e = Q_{el} + Q_{dyn} + Q_{add},$$
(II.1)

where Q_e are exploitable reserves; $Q_{el} + Q_{dyn}$, elastic and dynamic reserves; and Q_{add}, the reserves that are added during exploitation.*

In the Soviet Union, exploitable thermal water resources are subdivided into four categories--C_2, C_1, B, and A--ranked by increasing degree of exploration, research on water quantity, quality, and exploitation conditions (table 1). When a regional estimate of the exploitable reserves of geothermal water is prepared, the so-called hypothetical exploitable resources are determined. This term indicates a level of knowledge that is below the C_2 category.

In addition, exploitable reserves of thermal water are further subdivided into 2 groups, with respect to their current economic value:

(a) Reserves that can now be exploited economically, and which satisfy both the intended use and the proposed system of exploitation;

(b) Reserves that are presently uneconomical due to small volume of water for the intended purpose, or because they require extremely complex methods of exploitation; viewed as potentially exploitable in the future.

Classification is overseen by the State Commission for Mineral Reserves, which examines and approves for development all mineral, oil, gas, and subsurface water resources. The Commission is under the jurisdiction of the USSR Council of Ministers.

---

*During exploitation of a GWF, static reserves are not extracted.

TABLE 1

CLASSIFICATION OF EXPLOITABLE THERMAL WATER RESERVES
ON THE BASIS OF LEVEL OF STUDY

Category of Reserves	Characterization of Level of Study	Type of Reserves
C_2	Reserves have been established on the basis of geologic and hydrothermal data, and have been confirmed by sampling of water-bearing formations at specific locations, or by analogy with other explored areas.	Potential
C_1	Reserves have been explored and studied to the point where data assure general determination of the structure, stratigraphy, and extent of water-bearing formations. Exploitable reserves are determined by sampling data from single exploratory wells, and through analogy with well-known fields.	Potential
B	Reserves have been explored and studied in detail; data obtained assure determination of basic stratigraphic features, structure, and recharge of water-bearing formations. Indentification of recharge sources for replenishing exploitable resources of sub-surface water has been conducted. Water quality has been studied to the point where its use may be established. Exploitable reserves of subsurface water at site of projected supply have been determined on the basis of pump (flowing) tests and extrapolated estimates.	Commercial
A	Reserves have been explored and studied to the degree that data assure complete understanding of stratigraphy, structure, pressure heads, and permeability of water-bearing rock. Means by which water-bearing formations are replenished and their potential to add to exploitable reserves have been ascertained. Volume of subsurface water has been determined with certainty that assures the potential for desired purpose and useful life. Exploitable reserves of subsurface water have been determined at water supply site using pump (flowing) tests and/or other means.	Commercial

2. Drilling Operations Technology at Geothermal Sites

Boreholes are the principal means for exploring and developing GWFs. Generally, drilling technology and construction specifics for geothermal wells are determined by factors such as projected depth, bottomhole diameter, characteristics of the geologic section, purpose of drilled wells, and sampling technique. The selection of well construction techniques used further depends on whether the target water system is strato or fissure-vein type.

Drilling of Wells and Tapping of Thermal Water Under Stratal-Water Pressure Conditions

The geologic and hydrogeothermic conditions for drilling geothermal wells under confined conditions are similar to the conditions for drilling oil and gas wells. Often, testing of such wells results in significant thermal water flows. The average thermal water well depths are similar to those drilled for oil and gas as defined by artesian structures formed by sedimentary rock: usually, about 2,000 m. The bottomhole diameter for deep exploratory-prospecting wells in the USSR averages 152 mm. Deep exploratory and exploitation wells are drilled by the rotary or turbine methods; the projected depth of the well determines the drilling installation used (table 2).

When wells are drilled through sedimentary deposits (which often contain hydrocarbon deposits), blowouts of highly gasified water are possible. In order to prevent blowouts, the hydrostatic pressure of the drilling mud in the well is calculated so that it will exceed formation pressure by 5 to 15 percent, depending on the depth of the well. This is achieved by weighting down the mud by 2.2 to 2.4 g/cm^3 (usually with weighting compounds: hematite, magnetite, barite, or blast furnace dust

29

TABLE 2

SOVIET DRILLING INSTALLATIONS FOR DEEP WELLS

Type of Drilling Installation	Drilling Depth (meters)	Maximal Load Capacity on Grapple (tons)	Rotor Power (kW)	Maximal Pressure Produced by Pumps (kg/cm^2)	Overall Established Capacity of Main Drive (kW)	Installation Mass (tons)
BU-75	1,800	100	----	200	470.7	130
BU-80	2,800	140	220.65	200	993	257
Uralmash-125	4,000	160	367.75	200	1,655	308
Uralmash-125DG	4,200	200	367.75	250	1,265	495
Uralmash-125E	4,200	200	367.75	250		

having a density of 3.8-4.0 g/cm^3).* In addition, preventers are sometimes placed at the wellhead to facilitate sealing.

The structural features of deep geothermal wells must assure both safety and separate testing of the water-bearing formations. The following factors influence the achievement of these two goals: the number of casings placed into the well, the number of strings (also their external diameters and lengths), diameter of the stem per string, and the location of cementation intervals (depth of the upper and lower limits). Casing pipes used to reinforce oil, gas, and geothermal wells are generally manufactured to conform to GOST (USSR standard) 632-64. External diameters of pipes used in the USSR for oil wells range from 114 to 508 mm, and the lengths, 9.5 to 13 m.

When deep geothermal wells are reinforced, the following four types of casings are used:

(1) directional--to reinforce the upper interval that is characterized by unstable deposits and intended to prevent washout of the wellhead

(2) conductor--to reinforce the uppermost unstable rock, insulate water-bearing horizons from contamination, placement of anti-blowout equipment at the wellhead, as well as the support of subsequent casings

(3) interval casings--to reinforce and isolate upper-lying zones of the geologic section that are incompatible with the drilling procedures for the underlying strata and to prevent complications or accidents in the well during the drilling of subsequent intervals (under favorable conditions, intermediate casing is used as producing string)

(4) production string--to reinforce and separate strata that carry thermal water and to insulate them from other horizons in the geologic cut, as well as for production of thermal water.

---

*In standard conditions, density of drilling mud is maintained within the 1.18-1.22 g/cm^3 range.

Single-stage (one-cycle) cementing of wells is the most commonly used cementing technique. The materials used to cement wells include plugging cements that consist of binding substances (Portland cement, slag, lime, etc.), mineral substances (quartz sand, asbestos, clay, slag, etc.), or organic additives (cellulose production wastes, etc.) that provide a mud when mixed with water, and subsequently harden into cement. Special plugging cements are used for geothermal wells (GOST 158-63) with setting initiated no earlier than after 1 hour 15 minutes and durability no less than 18 kg/cm^2 after storage for 24 hours at a temperature of 100-120°C. As temperature and pressure are increased in the well stem, the setting time of the cement is decreased substantially; the setting time becomes so short that it is no longer possible to pump the cement mix into the annular space. Compounds are added to the cement mix as necessary to slow setting (salts of humin, ligosulfonic acids, or other surfactants).

Tapping of thermal water formations and outfitting the inflow part of the well constitute the most important stage in the drilling operations, since it is at this point that the degree of contact between the well stem and the stratum is established. The quality of the tapping of potential producing strata is dependent on the quality of data received during exploration of the well and the magnitude of water inflow. Technology for tapping horizons containing geothermal water is determined by filtration characteristics of the rock, formation pressure, temperature, gas saturation of the water, and other factors. If the hydrostatic pressure in the stratum is normal, the collecting strata have a tendency to absorb the drilling mud. In a geothermal well section (sand, stone, limestone), a great number of porous collector-strata can be found which are separated from one another by clay, argillites,

32

marl, dense sandstone, and other rock. Clay mud may permeate strata to dozens of meters in highly fissurated collectors, which acutely reduces permeability. In such cases subsequent measures for developing wells do not always yield positive results. Absorption of the drilling mud is also determined by the nature of the absorption site, and the pressure of the drilling mud in relation to formation pressure.

In order to effectively combat absorption, measures are carried out after reaching the absorption zone; these include geophysical and hydrodynamic procedures and taking of rock samples. Borehole gauging is carried out in the entire stem, and radioactive logging is performed in the interval being tested. Zones with extensive absorption and overflows are determined with flowmeters.

To prevent contamination of strata carrying thermal water during the drilling process, drilling mud parameters are usually changed so that the fluid has minimal density but at the same time has sufficient viscuous and thixotropic features (such as clay muds with surfactant additives). Another approach is to minimize the time lapse between penetration of the stratum and hydrogeologic testing procedures.

After the well is drilled to the planned depth and the thermal stratum is tapped, the water intake section of a well 1,500 m deep can be reinforced with liners of varying design, and the casing cemented.* In exploratory wells a number of geothermally promising sites are tested via sequential perforation of a preliminarily cemented string; perforation is carried out from bottom to top along the well stem. In

---

*The liner filters can be equipped with round as well as with slotted openings; however, such filters do not always prevent the entry of sand into the well and often plug. Metal-ceramic, sand-and-plastic screen, or gravel filters are used to prevent sanding up.

order to prevent a decrease in the collecting capacity of the test sites as a result of cementation of the annular space in the production string, the horizons are isolated by cementing the intervals between the objects designated for testing. After testing, the upper-lying horizons are isolated from those situated below and from one another by construction of cement bridges or packers in the string. Tapping is carried out by perforating the casing and the cement ring with bullet, torpedo, cumulative, or sand jet perforators. With cumulative perforation 8-9 mm holes are drilled and the penetration depth of the charges into the media (channel depth) may be up to 20-25 cm. The optimal perforation density is 8-10 openings 12-13 mm in diameter per meter of layer thickness. With the aid of a sand jet perforator, vertical or horizontal slots as well as direct channels can be achieved to 30 cm. However, sand jet perforation requires unwieldy above-ground equipment and costs more than cumulative perforation. The type and density of perforation is determined in each individual case depending on the collector features of the stratum, the construction of the well, temperature, and the estimated pressure in the perforation interval.

The last effort that precedes the testing of the well is the start of geothermal inflow. Cleaning of the bottomhole and start of inflow during development of geothermal wells are accomplished by flushing the well by, for example, injection of compressed air, or swabbing, or a combination of these methods. During flushing, clay fluid in the well is replaced with water; pressure at the bottomhole is reduced and the clay crust is removed. Flushing of wells is facilitated by an armature at the wellhead.

Drilling of Wells and Tapping Thermal Water under
Fissure-Vein Conditions

The techniques for drilling superheated geothermal wells in the
fissure-vein water pressure systems located in areas of contemporary
volcanic activity are influenced by changes in the state of the
superheated water during production and the high temperatures of the
cross-section. The depth of geothermal wells at fissure-vein GWFs
usually does not exceed 1,000 m (such wells are most frequently 400-600
m deep). These wells are drilled by the rotary method using ZIF-650A,
ZIF-1200A, and URB-4PM units equipped with powerful pump assemblies.

There are two principal features of drilling superheated water
sites: the well must be continuously and intensely flushed with a clay
mud at the rate of 20 l/s to cool the entire well stem during the
drilling process, and the clay drilling mud must be prevented from
boiling, which could result in a spontaneous steam-water mixture gusher.
When superheated water is tapped at a depth of several hundred meters,
the cooling of the well with intensive flushing guarantees prevention of
sudden steam-water mixture "explosions." It is considerably more difficult to prevent sudden steam-water blowout when superheated water is
tapped at a depth of several dozen meters (the flushing liquid column
does not balance the formation pressure in collectors with superheated
water). In such cases, it is expedient to drill geothermal wells outside the boundaries of natural discharge centers of superheated water.
Another important feature of drilling superheated water wells in areas
of contemporary volcanic activity is the need to provide effective
cooling of the circulating clay drilling mud, temperature of which often
reaches 70-80°C. Geothermal wells where steam-water production is
planned are equipped with reliable preventers which assure complete

sealing of the well stem, should it become necessary. The stem of such wells is reinforced with casing pipe, and the annular space is thoroughly cemented. The most effective diameter for the production string is 6 inches, and the walls of the well in the inflow interval of superheated water are reinforced with slotted or perforated pipes.

3. Direction of Geothermal Search Efforts

The search for GWFs--each of which is a unique combination of geologic, hydrogeothermic, geographic, and consequently, economic features, is usually conducted in areas that are not well explored. The study approaches applied to the search for geothermal water fields are based on the preliminary evaluation of geothermic, hydrogeologic, lithofacial, geostructural, and other prerequisite factors. The methodological plans for conducting exploratory efforts have been developed for GWFs that are limited to water pressure systems of the stratal and fissure-vein type.

Stratal Water Pressure Systems

The exploration of GWF stratal-water pressure systems is divided into two stages: 1) study, analysis, and generalization of geophysical, geologic, and hydrogeothermic data presented in the form of special maps and cross-sections, and compilation of geothermal zoning charts; 2) drilling and analysis of wildcat boreholes.

Stratal water pressure systems are usually confined to sedimentary formations that vary in lithology and dimension. In platform regions, stratal water pressure systems form large artesian basins that may extend millions of square kilometers (e.g., the West Siberian artesian basin). Artesian basins tied to foredeeps and intermontane depressions may extend tens of thousands of square kilometers (e.g., the Kura-

Araksinskiy artesian basin in Transcaucasia, etc.). In such structures, the thickness of the sedimentary deposits often reaches 10 km. Geothermal water in such artesian basins has a broad areal distribution and is found at a depth of more than 1,000 m, whereas the centers of recharge and/or discharge are usually located in faulted peripheral sections of foredeeps and intermontane depressions. In the USSR, stratal water pressure systems are widely distributed in West Siberia, the Caucasus, Central Asia, Kazakhstan, and elsewhere.

The viability of areas with a thermal water resource base is determined by a range of natural factors: the effective thickness and permeability of strata, temperature, mineralization level of the water, etc. Prospecting efforts for thermal water in stratal systems are carried out with a rather limited data base for the site's hydrogeothermic parameters. Geophysical surveys of artesian structures performed by surface techniques (seismic survey, electrical, gravity, and magnetic prospecting) establish only the general geologic and structural features of the search area. Additional data on geothermic and lithofacial characteristics in formations in deeply depressed basins are provided by a scattered network of stratigraphic information wells, and oil and gas wells. In order to evaluate individual water-bearing stratal complexes, the available geologic data are generalized in a series of schematic maps covering each formation: these maps show depth, geothermal, lithofacial, hydrochemical, and other factors. Maps plotted using subsurface mapping techniques characterize (at the first approximation level) the distribution of essential hydrogeothermic parameters for the individual water-bearing complexes.

The broad distribution of thermal water basins is accompanied by spatial non-uniformity, i.e., changes in the effective thickness of collectors, permeability, chemical composition and mineralization of stratal water, degree of heat, and other features. At the same time, typification of the natural hydrogeothermic environment (when stratal systems are being evaluated for geothermal potential) is to a significant degree complicated by spatial noncoincidence of structural, lithologic, and hydrogeologic boundaries.

The hypothetical exploitable reserves of thermal water serve as the overall indicator of the geothermal potential of a basin; this estimate is essentially based on the forecast of formation pressure dynamics under varying exploitation rates and over an extended time period. In the absence of detailed information on water-bearing complexes, it is legitimate to use simplified hydrodynamic stratal simulation models; these can then be applied in analytical functional relationships. The evaluation of these relationships shows that, all other conditions being equal, boundaries have a significant influence on drawdown in the stratum only when geothermal wells are placed comparatively close to these boundaries. Thus, when the distance from the geothermal water recharge area to the boundary of the stratum is equal to 25-30 km or more, the boundaries of the stratal water-pressure system have practically no influence on the behavior of the wells [18]. Experience with oil fields and industrial* and geothermal water fields also attests to the major influence of the spatial location of production areas (with respect to hydrodynamic boundaries) on the behavior of the stratum. For example, at formations with oil and geothermal water in the outcrop zone that outlines the northern slope of the Greater Caucasus, where recharge areas are found and where the piezometric head originates, wells have

---

*"Industrial" ground water includes water and brines from which valuable chemical elements or their compounds can be extracted.

been producing thermal water at a great rate for a long time, with individual wells producing 2,000 m^3/day (0.53 million gallons/day), e.g., at the Makhachkala field. These producing areas include the Tersko-Dagestan region, the Mineral'nyye Vody area in the Caucasus, and the Adygey anticline. At the Oktyabr'skoye Oil field near the city of Groznyy, over a lengthy period of exploitation (from 1916 to 1969) of the XII Karaganskiy horizon stratum (N_1), 259,555,604 t of water with an approximate temperature of 81°C have been extracted together with the oil [28]. This means that 13,400 m^3/day (3.5 million gal/day) of thermal water were produced. Long-term well production is possible when there is full compensation of extracted water by natural resources with an active water-drive regime. At the same time, when oil- and water-bearing beds are developed at sites situated at great distances from the recharge zone and at great depths from the surface (e.g., the Romashkin-skoye and Tuymazinskoye oil fields, Krasnodar Kray gas condensate fields, and the Krasnokamskoye, Chelekenskoye, Slavyansko-Troitskoye, and Neftechalinskoye industrial water fields), an elastic or elastic water-drive regime is produced, which has a decreased replacement coefficient (compared to that of an active water-drive regime).

The most important parameter for evaluating predicted exploitable thermal water resources on the basis of theoretical relationships is the filtration resistance factor*, which is included in all calculation formulas describing the flow dynamics of formation fluids into the wells. The magnitudes of the filtration resistance factors in pore (sandstone) and fissure-strata collectors (limestone, dolomite) that hold thermal water fluctuate rather broadly. For practical purposes strata with

---

*The filtration resistance factor is $\frac{\mu}{kh}$, or the reciprocal of the hydroconductivity factor $\frac{kh}{\mu}$; k is the permeability.

filtration resistance factors on the order of 2.5 · 10^{-4} cp*/darcy ·
cm and lower are of the greatest interest. At the Makhachkala and
Groznyy thermal fields, whose industrial values were determined as a result of detailed explorations, the filtration resistance factor of individual strata was in the 0.73 · 10^{-4} – 2.5 · 10^{-4} cp/darcy · cm range.

Temperature not only determines the magnitude of the thermal potential of the geothermal strata, but also has substantial influence on
thermal water dynamics. At 20°C, dynamic viscosity is 1 cp, and when
temperature rises to 100°C, water viscosity is gradually reduced to
0.282 cp (water mineralization has an inconsequential effect on the
viscosity factor). The reduction in viscosity of stratal water
increases with temperature and depth, and has a major effect on the
mobility of thermal water at great depths; at the same time, it partially compensates for the decreasing permeability of sand collectors,
which become more dense with depth. In order to unify the forecasting
of exploitable reserves on the basis of diagrammed hydrodynamic models
of stratal systems with the use of an analytical method, a hypothetical enlarged well with a regular flow calculated for a lengthy period of
time may serve as a unique indicator for forecasting reserves, or the
"water guarantee" factor [21]. By correlating this factor with maps
showing distribution of qualitative factors of thermal water basins
(temperature, depth, mineralization, etc.), a comparative analysis of
individual areas is carried out: subsequently, the field with the
required characteristics is selected.

In the mid-1960s, the principles of regionalization and evaluation
of stratal water-pressure systems illustrated above were utilized for

------

*cp is centipoise.

the first time in order to manage geothermal search operations in Eastern Precaucasia.* This region is an extensive artesian basin, covering an area of 150,000 km^2. The cross-section of this basin shows a series of water-bearing formations that appear on the surface of the northern slopes of the Caucasian Range (see figure 2). The hydrogeochemical zonality of subsurface water can be clearly demonstrated here, where subsurface waters travel from modern recharge zones toward the center of the basin. Starting with a depth of 1,000 m, water with a temperature range of 40°C and above may be widely distributed in the water-permeable strata. Using the principal water-bearing formations of the Eastern Precaucasian artesian basin, discrete hydrodynamic diagrams were compiled which, in combination with maps that characterized the distribution of qualitative geothermal factors, were used to carry out a comparative analysis of thermal water fields. Figure 3 shows the hydrodynamic regionalization of one of the most promising stratal water-pressure systems in Eastern Precaucasia--the Chokrakian water-bearing formation (Middle Miocene).

After laboratory work to determine promising geothermal fields is carried out and technical and economic feasibility studies are compiled, the second stage of exploratory operations, that of drilling exploratory wells, is initiated.

An example of the successful application of this technique for directing search operations in stratal systems is the discovery of the Kizlyar GWF in 1968 in Eastern Precaucasia. Drilling recommendations

---

*Precaucasia, or Ciscaucasia, is the part of the Caucasus north of the Caucasus Mountains that contains the Stavropolskiy and Krasnodarskiy areas (kraya) and the Dagestan, Chechen-Ingush, North-Osetin, and Kabardino-Balkar Autonomous Republics.

FIGURE 2

GEOLOGIC STRUCTURE AND RELIEF FOR THE

CENTRAL CAUCASIAN-PEREDOVYYE KHREBTY-ZATERECHNAYA PLAIN-ASTRAKHAN REGION PROFILE

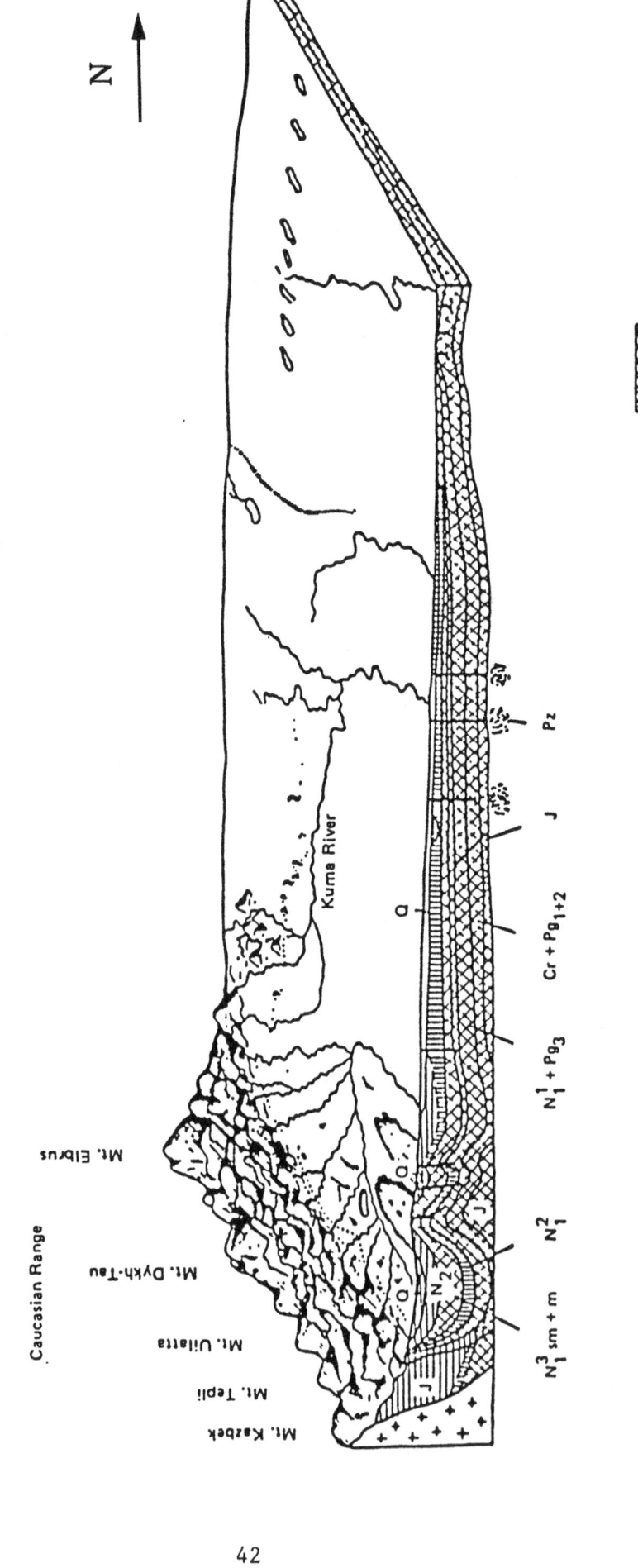

Key:

Sodium sulfate water

Sodium hydrocarbonate water

Calcium chloride water

Source: G. M. Sukharev, Gidrogeologiya neftyanykh i gazovykh mestorozhdeniy Publ. — Nedra, Moscow, 1971, p. 127

42

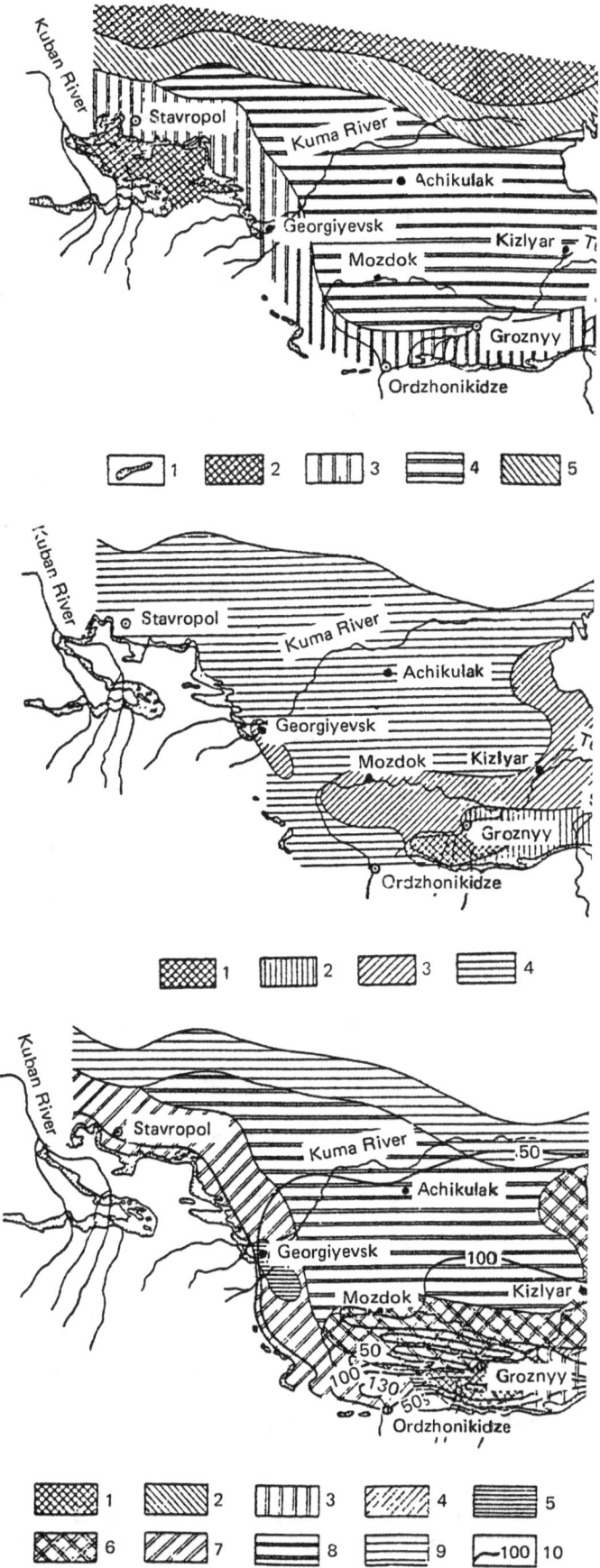

for a pioneer geothermal well at the Kizlyar field are given in reference 27.*

The high cost of drilling geothermal boreholes in stratal water-pressure conditions (because of the great depth of the water-bearing strata) determines the practical side of the exploratory effort: that of drilling single boreholes. At the same time, the areal distribution of geothermal water makes it possible to drill wells in the immediate vicinity where they will be used. The series of tests conducted at exploratory boreholes is essentially similar to the efforts conducted at pro-specting wells. These measures are described in part 4 of this chapter.

Fissure-Vein Water-Pressure Systems

Thermal water in fissure-vein type collectors is distributed in fold regions that have been affected by neotectonic activity (for example, the Caucasus, Pamirs, and Tien-Shan) and modern zones of volcanic activity (e.g., Kamchatka and the Kurile Islands). Fresh and brackish thermal water with temperatures to 200°C may be found in fissure-vein water-pressure systems located at a depth of less than 1,000 m. In these systems, it is difficult to establish recharge areas and pressure origins; however, discharge areas are usually characterized by surface thermal activity in the form of hot springs, geysers, fumaroles, hot steam areas, etc. Fissure-vein GWFs usually do not exceed several square kilometers. Due to the local occurrences of GWFs in fissure-vein systems, exploratory efforts are usually concentrated on these thermal anomalies. In addition, special measurements are taken and geophysical survey is used, which make it possible to determine and study faults,

---

*Identical techniques for estimating water-pressure systems were applied during exploration of other regions in the USSR as well [16].

44

dislocations, and areas of tectonic fracturing where thermal water circulates and is discharged. Since the end of the 1960s, infrared photography from planes and satellites has made it possible to locate areas with thermal anomalies.

Large-scale (1:10,000-1:25,000) hydrogeothermic and hydrochemical* surveys constitute a mandatory element of the exploration process that is carried out during the study of a fissure-vein GWF system. The comparative analysis of aerial, satellite, and geophysical surveys makes it possible to determine the most promising areas for geothermal exploratory efforts. The determination of hypothetical exploitable reserves in fissure-vein geothermal water systems, however, is usually conducted through empirical evaluations of data on the natural discharge. The thermal capacity of steam hydrothermae (steam-hot water sources) found in the volcanic regions of Kamchatka and the Kurile Islands (Pauzhetka, Bol'she-Bannoye, Zhirovskiye, Dolina Geyzerov, Goryachiy Plyazh, and other sources) is higher than the thermal capacity of water when temperature is the same. Determination of the calorific value of steam hydrothermae is carried out during the process of making hydrogeothermal measurements with the Polyak method [29]:

$$I_{swm} = i_s q + i_w(1 - q), \qquad (II.2)$$

where I_{swm} = the calorific value of the steam-water mixture that is nominally equal to its temperature prior to boiling; i_s = the calorific value of the steam at the given boiling temperature (when $100°C$ = 639 Kcal/kg); i_w = the same, for water (with $100°C$ equal to 100 Kcal/kg);

---

*A characteristic chemical component of thermal water is the hydrosilicate ion, $HSiO_3^-$, since silica acid is better dissolved in thermal and alkali water.

q = the content of steam in 1 kg of steam-water mixture; $(1 - q)$ = the content of water in 1 kg of steam-water mixture.

To determine the content of steam in the boiling source, steam separation is suppressed with cold water in which steam will condense. Based on the flow rate, water temperature, content of ion chlorides prior to suppression with cold water (corresponding to Q_{hot}, T_{hot}, and C_{hot}), identical data for cold water fed into the gryphon (Q_{cold}, T_{cold}, and C_{cold}), and mixed water that flows out of the gryphon after flooding (Q_{mw}, T_{mw}, and C_{mw}), the steam content of the steam-water mixture is calculated by three different methods:

a) the volume method   (II.3)

$$q = \frac{Q_{steam}}{Q_{swm}} = \frac{Q_{steam}}{Q_{steam} + Q_{hot}} = \frac{Q_{mw} - Q_{cold} - Q_{hot}}{Q_{mw} - Q_{cold}} \ ;$$

b) the temperature method

$$T_{swm} = I_{swm} = \frac{Q_{mw} T_{mw} - Q_{cold} T_{cold}}{Q_{mw} - Q_{cold}} \ ; \qquad (II.4)$$

c) the chemical method

$$q = \frac{C_{hot} - C_{swm}}{C_{hot}} = \frac{C_{hot} - \dfrac{Q_{mw}C_{mw} - Q_{cold}C_{cold}}{Q_{mw} - Q_{cold}}}{C_{hot}}$$

$$(II.5)$$

$$= 1 - \frac{Q_{mw}C_{mw} - Q_{cold}C_{cold}}{(Q_{mw} - Q_{cold})C_{hot}} \ .$$

The concentration of chloride ion in the steam is taken to be zero. As a result of several tests using this technique, the thermal content of steam-water mixture at the Bol'she Bannoye field was determined to be

140-200 Kcal/kg. Later, during a detailed exploration of this field by the Kamchatka Geologic Administration, identical efforts were performed and the heat level at the steam hydrothermae was established at 140-170 Kcal/kg. At the Pauzhetka field, steam-water mixture heat value at natural discharge points was 140-150, and averaged 170 Kcal/kg inside the wells. In Kamchatka's Geyser Valley, steam hydrothermae attain temperatures of 250 Kcal/kg. Discovery of a hidden discharge of thermal water may be made on the basis of hydrochemical measurements because rising hydrothermae often discharge under surface bodies of water and streams or into unconfined acquifers. For example, it was established that the visible flow of Middle Paratunka springs does not exceed 30-40 1/s. When the Middle Paratunka area was drilled, a stable flow in excess of 80 1/s and a temperature of 83-86°C was produced from four wells alone. In the opinion of one of the most authoritative researchers of Kamchatka's hydrothermae, V. V. Averiyev, the overall flow of wells placed at the discharge source of hydrothermae is substantially greater than the visible discharge in natural conditions.

A stratigraphic profile of a fissure-vein GWF usually consists of effusive, intrusive, and sedimentary-metamorphic rock, which is characterized by great variability in petrographic and lithologic composition, structure, texture, level and type of fissuration, etc. These factors determine the broad range of changes in aquaphysical and mechanical properties of such sections, and consequently, significantly complicate the interpretation of geophysical surveys. The accumulated data on fissure-vein GWFs show that electro-surveys and seismic surveys are the most favorable methods. Good results in establishing tectonic fault areas have been provided by direct current electro-profiling methods, espe-

cially the bilateral dipolar profiling method (DP)*. The most widely
used seismic survey method is KMPV** (correlative refractive wave profiling method), which makes it possible to locate fissure zones.

Large tectonic faults that extend to a great depth and which are
long as well as broad play a substantial role in the formation of
GWFs. Such faults can conduct thermal water from below, and the zones
they are found in are usually characterized by an abundance of water.
Prospecting boreholes are usually placed along such zones—these zones
have been determined using the series of photographic and geophysical
survey data, as well as research that was conducted earlier.***

4. <u>Exploratory Efforts at Geothermal Fields</u>

The exploration (development) of a GWF consists of a series of
drilling efforts and well studies that make it possible to determine the
spatial location and the geologic structure of the field. Exploration
also seeks to determine hydrogeothermic and technological parameters and
evaluate the commercial reserves. GWFs are examined as an overall use-
ful resource, one that has value not only as a source of thermal energy,
but also as a source of medicinal water and a potential source of raw
material for industry.

The types of analysis conducted at pioneer and exploratory boreholes can be divided into four groups: (1) observations made during
drilling of the well, (2) field-geophysical investigations (logging),
(3) flowing tests of thermal water and water-steam mixture from bore-

---

*<u>Dipol'noye profilirovaniye</u>

**<u>Korrelyatsionyy metod prelomlennykh voln</u>

***The entire territory of the USSR has been surveyed by the State Geo-
logic Survey using a 1:200,000 scale; some regions have been surveyed
using a larger scale.

holes, and (4) hydrochemical and technological evaluations of the thermal water. Research methodologies at such sites differ, taking into account the geologic, structural, and hydrogeothermic features of GWFs in stratal and fissure-vein water-pressure systems.

Stratal Water-Pressure Systems with Geothermal Water

Geothermal water of stratal water-pressure systems of average potential is usually used for heating and providing hot water supply for residential, industrial, and agricultural uses. In a number of cases, the wide distribution of thermal water basins in stratal systems makes it possible to drill exploratory-production wells in the immediate vicinity of the user facility. Well location takes into account the structural features of the producing strata, the degree of well interference, and the location of the user.

The number of exploratory-production wells required to explore a GWF is determined on the basis of preliminary evaluation of the field. When the exploitable resources are ranked in commercial categories, the overall actual flow rate of exploratory-producing wells must represent 50 percent of the required quantity of thermal water.

The selection of the thermal water-producing strata is made on the results of a series of studies and observations conducted during the well-drilling process. The measures include the study of core samples and cuttings, noting the balance and physicochemical composition of the drilling mud, logging, and analysis of individual section intervals with drill stem testing.

During drilling, continuous analysis of the cuttings provides additional data on the lithology and observation of the balance and quality of drilling mud often makes it possible to make a preliminary evaluation of the bedding interval of the water-bearing horizon and the chemical composition of the stratal water.

Logging of geothermal wells is one of the most important means for evaluating the geologic cross-section and other features of a GWF. There are approximately 40 different logging methods. The mandatory types of logging carried out at all geothermal wells are the standard electro-logging (AR and SP*), radioactive logging (GRL and NGRL**), and thermal logging. Other types of logging are included in the set of geophysical borehole-type operations, depending on the geologic and hydrogeologic characteristics of the rock (table 3).

Quantitative evaluation of water-bearing areas determined during the process of drilling wells is sometimes carried out with the aid of a drill stem test. This significantly reduces the overall number of sampling intervals and expenditures. For example, in order to examine strata in geothermal wells without casings, bottomhole-anchored formation testers are lowered into the well on pipes (IPG GrozNII, KII GrozUFNII); anchorless formation testers are also used (with IP-8 pipe).

Flow tests are carried out on flowing wells in order to determine the hydrodynamic parameters of the strata and wells. This information is necessary in order to evaluate the exploitable reserves of geothermal water and to estimate its technological properties and the commercial value. In accordance with the basic steps of geologic and exploratory efforts for the study and evaluation of a GWF the flow tests are subdivided into trial, experimental, and experimental-production procedures.

---

*AR - apparent resistivity; SP - self-potential

**GRL - gamma-ray log; NGRL - neutron-gamma ray log

TABLE 3

EXAMPLES OF BASIC BOREHOLE-GEOPHYSICAL LOG METHODS USED TO STUDY GEOLOGIC AND HYDROGEOLOGIC CONDITIONS

Type of Section	Rock and Measurement Conditions	Electric Logging		Radioactive Logging		Other Studies	
		Technique	Problems Being Solved	Technique	Problems Being Solved	Technique	Problems Being Solved
1	2	3	4	5	6	7	8
Sand-clay	Sandstones of varying porosity and grade, separated by clay strata of average and large thickness. Mineralization of stratal waters is relatively low; it changes appreciably along area and section; reduced penetration of the solution possible	Standard logging; measurement of AR and SP; measurement of curves with standard microlog sonde; in the producing area, with LLS	Correlation; dissection of section; detection of sand collectors and evaluation of their collector and water-bearing properties; determination of effective capacity; determination of mineralization of stratal water with SP	GRL, NGRL	Dissection and correlation of section; evaluation of rock porosity	Caliper log	Dissection of section; more precise definition of lithology; detection of collectors; well diameter data used for interpretation of geophysical curves
	Same; high mineralization of stratal water, changing little along area and section; higher penetration of solution	Same	Same; stratal water mineralization not measured	Same	Same	Same	Same
	Highly clayey section; collectors are formed from fine-grained materials (sand-aleurite strata) and are very porous	Same	Correlation; dissection of section; detection of sand-aleurite strata; evaluation of collector and water-bearing properties; determination of effective capacity	Same	Same	Same	Same

Key:

AR   - Apparent resistivity
GGRL - Gamma-gamma ray log
GRL - Gamma-ray log
LLS - Lateral logging sound
NGRL - Neutron-gamma ray log
SP   - Self-potential

51

TABLE 3, CONTINUED

Type of Section	Rock and Measurement Conditions	Electric Logging		Radioactive Logging		Other Studies	
		Technique	Problems Being Solved	Technique	Problems Being Solved	Technique	Problems Being Solved
1	2	3	4	5	6	7	8
	Carbonate rock of varying porosity and clay rock with relatively low resistance	Standard logging; measure AR and SP; measure AR curve with standard microlog sound; in producing LLS area or lateral logging	Same	Same	Dissection and correlation of cut on basis of GRL; quantitative evaluation of clay level in rock	Same	Same; well diameter increase is balanced against highly clayey carbonate rock
	Highly porous carbonate strata with intergranular porosity; clay rock	Standard logging; measurement of AR and SP; measurement of AR curve with standard microlog sound, in producing LLS area or lateral logging	Correlation; dissection of cut; detection of highly porous collectors, evaluation of collector and water-bearing properties; determine effective capacity	GRL, NGRL	Dissection and correlation of cut; quantitative evaluation of clay rock; isolation of highly porous granular collectors (using electrologging data)	Record caliper log using large horizontal scale	Dissection of cut into highly porous and low porosity rock with inter granular porosity; well diameter data used for interpretation of caliper
Hydro-Chemical Sediments	Anhydrite, gypsum, rock salt (halite, sylvite)	Same	Correlation; dissection of section	Same; augmentation of GGRL diagrams	Dissection and correlation of section; GRL curve used to evaluate clayiness and secretion of KCl salt-sylvite; GGRL curve for secretion of salt-halite and sylvite; NGK curves for anhydride and gypsum strata	Acoustic logging	Dissection of section; definition of lithology; detection of fractured collectors and evaluation of collector properties
	Sand strata have very low resistance--less than 0.5 ohm; increased penetration of solution	Standard logging; measurement of AR and SP; measure AR with standard microlog sondes; induction logging in the producing section	Correlation; dissection of section; detection of sand-aleurite strata and evaluation of their collector and water-bearing properties; determination of effective capacity	Same	Same	Same	Same

TABLE 3, CONTINUED

Type of Section	Rock and Measurement Conditions	Electric Logging		Radioactive Logging		Other Studies	
		Technique	Problems Being Solved	Technique	Problems Being Solved	Technique	Problems Being Solved
1	2	3	4	5	6	7	8
	SP curve is imprecise and of low quality; drilling mud is highly mineralized	Standard logging, measurement of AR and SP; lateral logging and microlaterolog survey	Correlation; dissection of section; detection of sand-aleurite strata and evaluation of their collector and water-bearing properties; determination of effective capacity	GK, NGK	Dissection and correlation of section; evaluation of clayiness of rock	Caliper log	Dissection of section; definition of lithology; detection of collectors well diameter data used for interpretation of geophysical curves
	Sand-clay section containing low-porosity strata that are cemented	Standard logging; measurement of AR and SP; measurement of AR curves with standard microlog sondes, in the LLS producing section or lateral logging	Correlation; dissection of section; detection of collectors and evaluation of their water-bearing properties; determination of effective capacity	Same	Gamma-logging dissection, correlation of cut and evaluation of clayiness of rock; detection of low-porosity rock with NGRL	Same	
Carbonate	Low-clay carbonate rock, fractured and fractured-cavernous with high resistance	Standard logging; measurement of AR and SP, lateral logging in the producing area; repeat lateral logging with varying resistance of drilling mud	Correlation; dissection of section; detection of collectors and evaluation of their water-bearing properties; determination of effective capacity	Same	Dissection and correlation of section; GRL quantitative evaluation of rock clayiness; NGRL and electro-logging used to detect collectors their porosity and water-bearing capacity	Record caliper log using large horizontal scale	Detection of intervals with greater fracturing and cavernosity; well diameter data used for interpretation of other log surveys

Source: [30].

Hydrogeological observation and procedures conducted during different kinds of flow tests do not vary greatly from test to test. What difference there is consists in the duration of the procedures and the number of wells tested. Trial flow tests are performed with single (pioneer) wells, and usually take several days. All types of flow tests are carried out from exploratory wells. Flow rate, pressure, water temperature, and mineralization are the parameters that are tested; also noted are pressure at wellhead of other exploratory wells (to evaluate the magnitude of the depression cone). As a rule, experimental flow tests are conducted at 2-3 different flow rates, and the overall duration may be from 10 to 30 days. Experimental-production tests are conducted at wells to study well interference and the rate at which formation pressure changes over time. Such tests represent a time-and-space fragment of the projected production of the GWF, and their duration may last for several months. During the flow tests, the chemical composition of underground water is measured. Experimental-production tests make it possible to produce reliable data in order to extrapolate the behavior of the GWF during extended exploitation (for a 25-year period).

In a number of cases, type MGG or MGP bottomhole manometers (with a sensitivity threshhold of 0.2-0.6 percent from pressure measurement limit) used to measure small pressure differentials of several atmospheres during flow tests were found to be unsuitable because measurement of depression accuracy could amount to 30-40 percent.* More accurate

---

*Excess static pressures at geothermal wells of the Makhachkala, Ternair, and Groznyy (Khankal'skaya valley) GWFs were 3-6 atm, while during thermal water overflow, it was necessary to measure depressions on the order of 1 atm.

differential manometers were not yet widely used in hydrogeologic practice at that time due to complexity and inadequacy of in-service use. Also, direct measurements of wellhead pressure at deep geothermal wells, even with the aid of precision manometers, did not make it possible to determine the true value of shut-in (static) and flowing (dynamic) wellhead pressures. Instances have been known when geothermal wells with a static level below that of the ground surface continued to overflow spontaneously after short-term forced pumping had been initiated (e.g., Sukhumi, Yevpatoriya). Shut-in and flowing wellhead pressure at deep geothermal wells depends not only on hydrodynamic factors (pressure distribution in the stratum, flow rate of well, and reservoir properties of water-bearing strata), but also on temperature changes of water in the well stem. This phenomenon is called "thermolift" and is connected with changes in the linear dimensions of the water column in the borehole when sharp temperature fluctuations occur during the testing process. This makes it necessary to process wellhead measurements taken at flowing geothermal wells in a special manner. To determine the true reservoir and producing bottomhole pressures using monometric wellhead measurement data, a calculation method based on the study of thermophysical properties of stratal waters is used [20, 22].

Determination of Static and Dynamic Pressures
of Deep Geothermal Wells Using Wellhead Data

In a geothermal well, water density depends on the temperature distribution along the well stem: from the wellhead to the bottomhole, water density gradually decreases. To determine water density depending on temperature when $t > 4°C$ the following formula can be used:

$$\gamma_{t_2} = \gamma_{t_1} - \alpha(t_2 - t_1) \quad , \qquad \qquad \text{(II.6)}$$

where γ_{t_2} = the density (specific weight) of stratal water when temperature is t_2, °C; γ_{t_1} = the density of strata water at temperature t_1, °C; α = the temperature density coefficient, g/cm^3 · °C.

For fresh water and slightly mineralized water, when heated, $t_2 >$ t_1, $\gamma_{t_2} < \gamma_{t_1}$, and when cooled, $t_2 < t_1$, $\gamma_{t_2} > \gamma_{t_1}$. Density of distilled water in a temperature range 4 to 100°C decreases from 1 to 0.958382 g/cm^3, and the value of the temperature-density coefficient of fresh water in a temperature range of 4 to 100°C, $\alpha \approx 4.16 \cdot 10^{-4}$g/cm^3 · °C. After substitution of this value in formula II.6, the following equation is derived:

$$\gamma_t = 1.001664 - 0.000416\, t \quad , \qquad \qquad \text{(II.7)}$$

where γ_t = the density of water in g/cm^3 at temperature of t, °C.

Formula II.7 is correct at $t > 4$°C and provides accuracy to 1%, because relationship $\gamma_t = f(t,$ °C$)$ is nonlinear. The value of fresh water density may also be determined using special tables. To a large degree, highly mineralized subsurface water contains a mixture of sodium chloride and calcium chloride (the former usually predominates), and at the same temperatures, density of this water is substantially different from fresh water. For this reason, formula II.7 cannot be used when mineralization exceeds 10 g/l. The value of the temperature-density coefficient, α, for distilled water and water with NaCl can be determined using the diagram in figure 4.

CHANGES OF TEMPERATURE-DENSITY COEFFICIENT OF DISTILLED WATER AND NaCl SOLUTIONS DEPENDING ON TEMPERATURE LEVELS

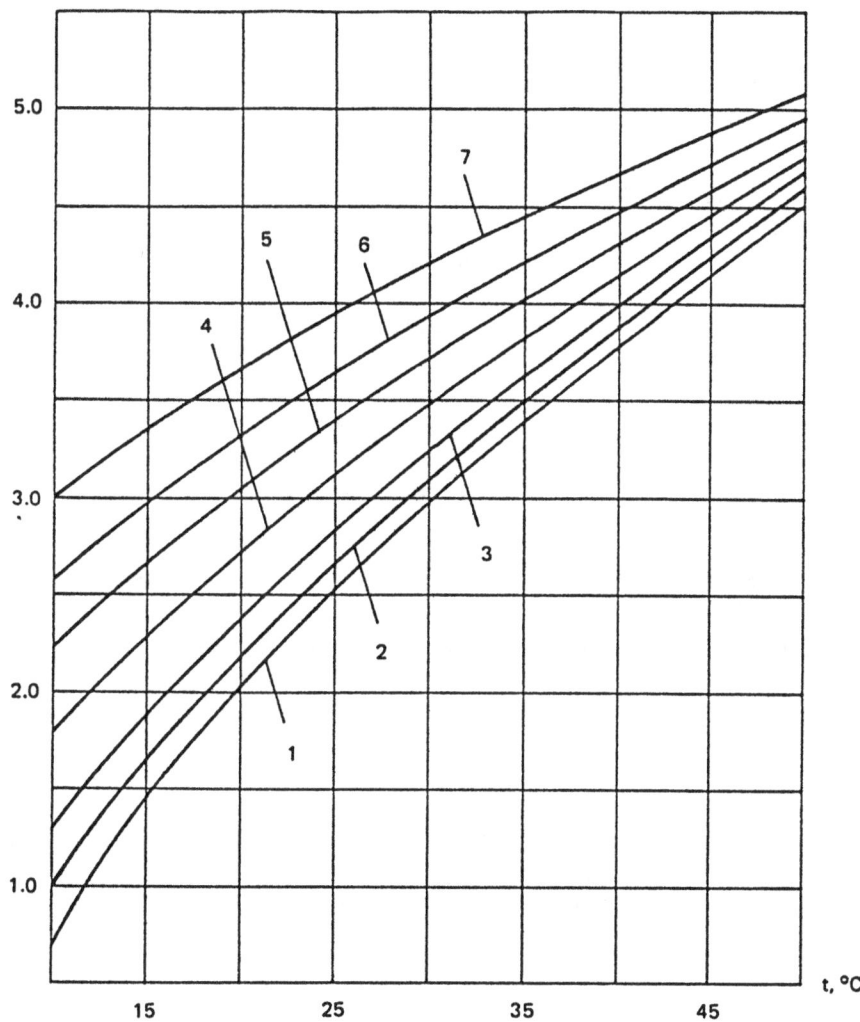

Water mineralization, g/l:

1. 0
2. 10.05
3. 20.25
4. 41.07
5. 62.47
6. 84.47
7. 107.07

Source: Polevoy, S. L., V. I. Savchenko, and V. V. Lisovin.
         "Opredeleniye gidrodinamicheskikh parametrov po dannim oprobo-
         vaniya skvazhin." Razvedka i okhrana nedr. Moscow, 1972,
         10:46.

Water column pressure in a temporarily shut-in well (formation or reservoir pressure) can be determined by the equation

$$P_r = P_{ms} + \int_0^H \gamma(H)dH \quad , \qquad (II.8)$$

where P_r = the formation pressure; P_{ms} = manometric (excess) static pressure at wellhead; H = the length of the water column in the well.

For practical calculations, it is assumed

$$\int_0^H \gamma(H)dH \approx \sum_{i=1}^{i=n} \gamma_i \Delta H_i \quad . \qquad (II.9)$$

In a deep geothermal well that is shut-in for an extended period of time, the temperature along the stem is distributed in accordance with the geothermic conditions of the stratal section. If density (γ_i) of portion i of water column in the stem is expressed using the right side of equation II.6, and the temperature distribution in the stem of the shut-in well using a graduated thermogram is expressed as

$$t_i = t_0 + \sum_{i=1}^{i=n} G_i H_i \quad , \qquad (II.10)$$

then a formula can be derived to determine the formation pressure:

$$P_r = P_{ms} + 0.1 \sum_{i=1}^{i=n} \left\{ \gamma_{i-1} - \alpha\left[(t_0 + \sum_{i=1}^{i=n} G_i \Delta H_i) - t_{i-1}\right] \right\}_i \Delta H_i \,[atm] \quad , \qquad (II.11)$$

where P_r = formation pressure, atm; P_{ms} = manometric static pressure at wellhead, atm; n = the number of rectilinear sections in the thermogram; ΔH_i = the projected length of rectilinear section of the temperature log with the number i on the ordinate, m; t_i = the temperature at end of portion i of water column in well, °C; t_o = the temperature of layer

with constant temperature, °C*; $G = \dfrac{\Delta t_i}{\Delta H_i}$, the geothermic gradient for

the thermographed section i, in °C/m.

To calculate the formation pressure in geothermal wells with fresh

and slightly mineralized water, formulas II.7 and II.11 are used to

derive the relationship

(II.12)

$$P_r = P_{ms} + 0.1 \sum_{i=1}^{i=n} \left[1.001664 - 0.000416 \left(t_o + \sum_{i=1}^{i=n} G_i \Delta H_i \right) \right]_i \Delta H_i \; [\text{atm}] \quad .$$

As a rule, measurement of static pressure at the wellhead of

geothermal wells is carried out when the well is tested for inflow

(after trial or experimental flowing test). In such cases, temperature

distribution along the well stem does not correspond to the natural tem-

perature distribution along the geologic section and often the relation-

ship t = f(H) approximates that of the linear relationship (figure 5).

To determine distribution density of water in the well stems of

wells shut-in for a short period of time, it is necessary to obtain a

temperature log immediately after measuring the excess pressure.

When the geothermal well is flowing, temperature is distributed

along its stem in a rectilinear, or nearly rectilinear fashion. Taking

into account pressure losses achieved as the water flows from the well,

bottomhole pressure may be calculated using the following formula:

---

*The temperature of a layer with constant temperature (neutral layer)
usually assumes average long-term temperature of the air in the area of
investigation.

FIGURE 5

TEMPERATURE SURVEYS FOR GEOTHERMAL WELLS AT THE MAKHACHKALA GWF

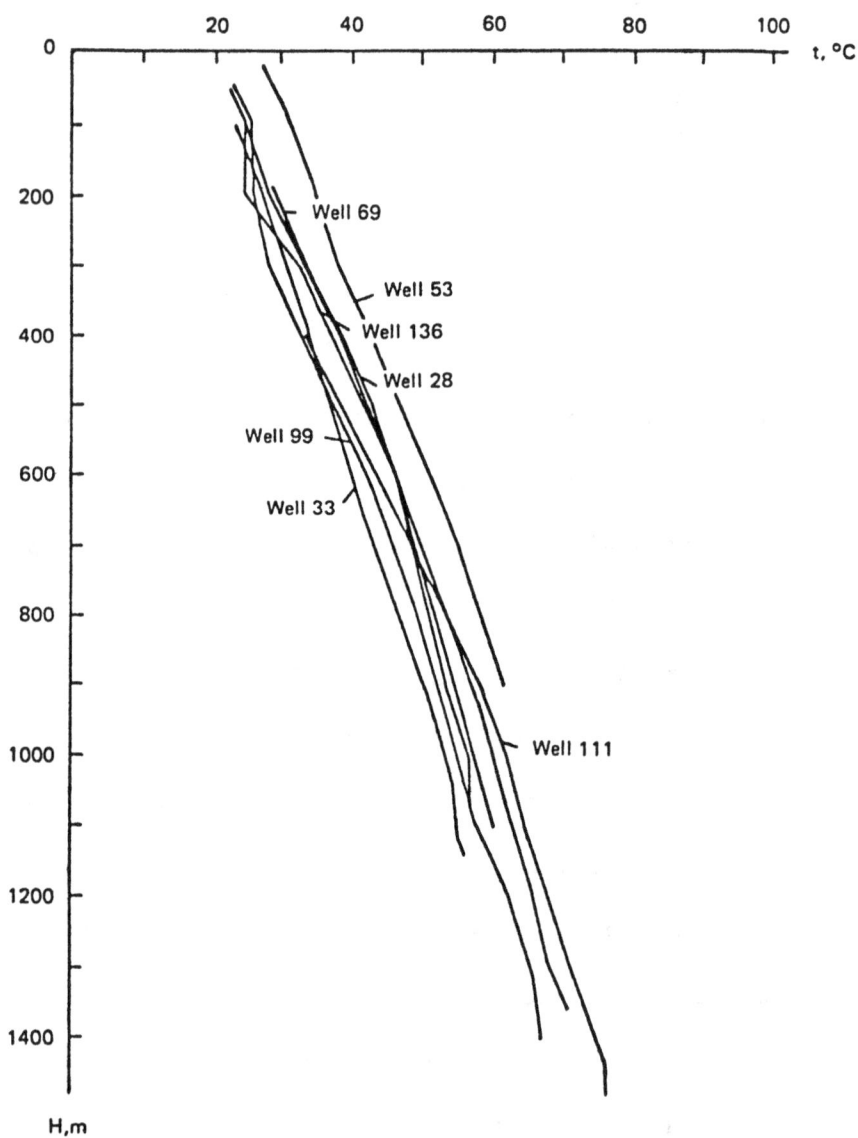

Source: Polevoy, S. L., I. P. Kovshov, and I. L. Lumelskiy.   "Obrabotka
          rezul'tatov opytnykh vypuskov termal'nykh vod iz skvazhin."
          In: <u>Razvedka i okhrana nedr</u>.   Moscow, 1968, 11:51.

$$P_{fb} = P_{md} + P_1 + 0.1H\left\{\gamma_{t_{fh}} - \alpha\left[\left(\frac{t_r + t_{fh}}{2}\right) - t_{fh}\right]\right\} [\text{atm}] \quad , \quad (\text{II.13})$$

where P_{fb} = the bottomhole flowing pressure, atm; P_{md} = the flowing wellhead (manometric) dynamic pressure, atm; P_1 = pressure losses caused by friction as water flows from bottomhole to the wellhead, atm; t_r = reservoir temperature, °C; t_{fh} = the water temperature at the wellhead of the flowing geothermal well, °C; $\gamma_{t_{fh}}$ = density of the water at t_{fh}, g/cm3.

Pressure losses in wells can be significant. Experimental efforts carried out at one of the 1,670 m-deep injection wells at the Romashkino Oil field (water production 390 m3/day, or about 100,000 gallons per day, with pipe diameter, d = 2"), determined that pressure losses reached 36.1 atm [31]. These losses can be calculated with the Darcy-Weisbach formula. Accuracy in determining hydraulic losses in the well stem is determined primarily by the value of the hydraulic resistance, or friction coefficient, λ. Calculation formulas for determining this value include the Reynolds number, Re, and take into account the roughness of the well walls.

To determine pressure losses due to flow of water along the well stem, an experimental formula of the Moscow Petroleum Institute can be used [31]:

$$P_1 = 82.6\,\lambda\,\frac{\gamma Q^2}{d_{in}^5}\,L \quad , \quad (\text{II.14})$$

where P_1 is pressure loss, atm; Q = the water flow, l/s; d_{in} = the inner diameter of the well, cm; γ = the specific fluid weight, g/cm3;

61

L = the depth of the well, m; λ = the hydraulic resistance coefficient equal to 0.02-0.015.

During tests of geothermal wells at the Makhachkala GWF, dynamic pressures taken at the wellhead exceeded static pressures in several instances; this was caused by the thermolift phenomenon. Indicator diagrams, $\Delta P* = f(Q)$, plotted using data for geothermal well pressure, had the shape of disjointed lines and were unsuitable for calculation of the stratum filtration parameters (figure.6). The scattering of points on the indicator lines diagram is related to the simultaneous effect of thermolift and hydraulic losses in the well stem, which result in changes of wellhead pressure, with varying absolute value and sign: pressure changes due to the thermolift phenomenon have a positive sign, and those which result from hydraulic friction losses, a negative sign. Generally, regardless of the nature of temperature distribution along the well stem and the water mineralization level, the following formula can be used to calculate pressure depression:

(II.15)

$$\Delta P = 0.1 \sum_{n=1}^{i=n} \gamma_i \Delta H_i - 0.1 H \left\{ \gamma_{t_{fh}} - \alpha \left[\left(\frac{t_r + t_{fh}}{2} \right) - t_{fh} \right] \right\}$$

$$+ P_{ms} - (P_{md} + P_l) [atm] \quad .$$

When temperature distribution in the well stem is linear, the following formula is applicable to calculate the drop in formation pressure:

(II.16)

$$\Delta P = P_{ms} - P_{md} + \left[0.05 \, \alpha H (t_{fh} - t_{red}) - P_l \right] [atm] \quad ,$$

where t_{red} = reduced temperature, $t_{red} \approx t_o$ (t_o = the temperature at the stratum of zero temperature amplitude). On the basis of formula II.16, results of flow tests at Makhachkala and Ternair GWFs were calculated.

DIAGRAM $\Delta P = f(Q)$ FOR MAKHACHKALA WELL 24-T

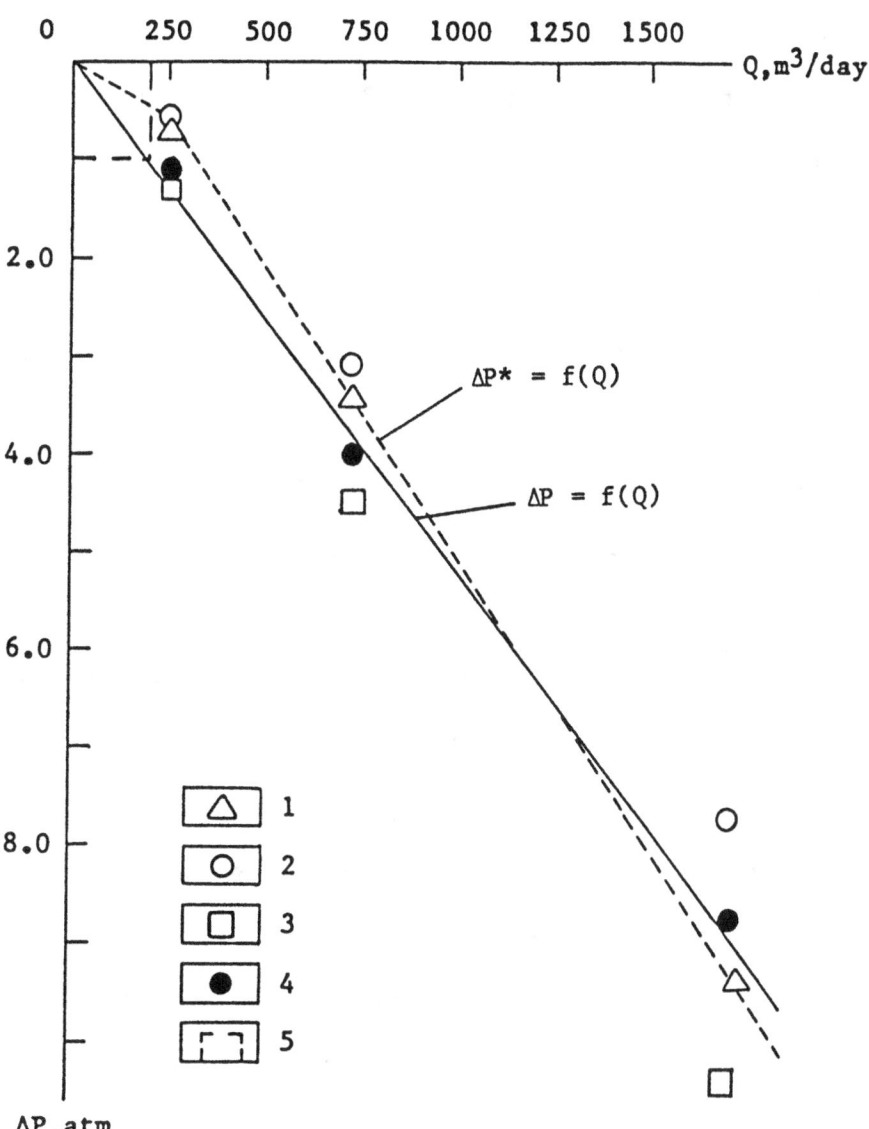

Key:

1 - points for measured values of depression, ΔP^*

2 - points that take into account corrections for pressure losses in the well, $\Delta P^* - \Delta P_{fr}$

3 - points that take into account corrections for thermolift, $\Delta P^* + \Delta P_{tl}$

4 - points that take into account corrections for losses of pressure and thermolift, $\Delta P^* - (\Delta P_{fr} - \Delta P_{tl})$

5 - calculated specific rate of flow of when $\Delta P = 1$ atm, q = 190 m^3/day

Source: Polevoy, S. L., V. I. Savchenko, and V. V. Lisovin.
         "Opredeleniye gidrodinamicheskikh parametrov po dannim oprobo-
         vaniya skvazhin." In: Razvedka i okhrana nedr. Moscow, 1968,
         11:44.

Figure 6 shows indicator lines for well 24-T (Makhachkala). The well was set into the Karaganian water-bearing horizon, 1,378 m deep; diameter, 6"; bottomhole temperature, 62°C; t_{red} = 20°C; overall water mineralization, 4.8 g/l; and manometer static pressure in shut-in well, P_{ms} = 9.7 atm. Testing and depression calculation results for this well are shown in table 4. The indicator line for this well, plotted along calculated depressions, has a shape that approximates a straight line (figure 6). The diagram $\Delta P = f(Q)$ makes it possible to determine the specific flow rate of a well; this rate can then be used to calculate the filtration parameters using formulas for the steady-state fluid inflow into the well.

Deep geothermal wells are exploited under difficult conditions compared to potable-water wells. Among the phenomena that complicate the operation of these wells are sand brought up from the bottomhole area, the formation of sand bridges, pipe corrosion, and salt crystallization in the production casing and wellhead armature.

During flow tests of geothermal water, detailed analysis is also performed on chemical composition for the presence, quantity, and composition of dissolved gas, and for the presence of harmful additives. For this purpose, water samples are taken from wells for field, partial, or complete analysis. Special water samples are also taken to analyze the content of dissolved gases. The definitions for each type of analysis are presented in table 5. When complete analyses are performed, rare and trace elements such as Li, Sr, Ba, B, Al, Mn, Co, As, Br, Rb, Cs, Zn, Pb, and W are found. In addition, the presence of organic compounds is determined.

Analysis of dissolved gases consists in determining presence of hydrogen sulfide (H_2S), carbonic acid (CO_2), O_2, CH_4, heavy hydrocar-

TABLE 4

TESTING AND DEPRESSION CALCULATION RESULTS FOR MAKHACHKALA WELL 24-T

| Well Output (Q, m³/day) | Wellhead Dynamic Pressure (P_{dyn}, atm) | Wellhead Water Temperature (t, °C) | Measured Depression (ΔP^*, atm) | Corrections, atm | | | Calculated Depression (ΔP, atm) |
				For Friction (P_{fr})	For Thermolift (P_{tl})	$P_{fr} + P_{tl}$	
254	9.3	51	0.4	−0.04	+0.79	+0.75	1.15
720	6.3	57	3.4	−0.28	+0.95	+0.67	4.07
1,728	0.36	59	9.34	−1.60	+1.01	−0.59	8.75

TABLE 5

TYPES OF WATER ANALYSIS AND REQUIRED DATA

Type of Analysis	Required Data	Description
Field	Physical properties; alkalinity; Cl^-	Carried out under field conditions. Used for preliminary characterization of water in the region and content of useful elements. No verification of test.
Partial	Physical properties; pH; CO_3^{2-}; HCO_3^-; Ca^{2+}; Mg^{2+}; SO_4^{2-}; Cl^-; Na^+ + K^+ (according to differences); hardness (general, carbonate, non-carbonate); CO_2 corrosion; mineralization; dry residue	Carried out in laboratory conditions. Used for mass characterization of the region's water and content of useful elements. Verified on the basis of dry residue.
Complete	Physical properties; pH; Eh; NH_4; Fe^{2+}; Fe^{3+}; NO_3^-; NO_2; acidity; H_2S; CO_3^{2-}; HCO_3^-; Ca^{2+}; Mg^{2+}; Na^+; K^+; SO_4^{2-}; Cl^-; SiO_2; dry residue; hardness (general, carbonate, non-carbonate); CO_2 corrosion; mineralization	Carried out in laboratory conditions. Used for detailed characterization of the region's water and content of useful elements. Verified on the basis of dry residue and sum total of cations and anions.

66

bons, nitrogen (N_2), Kr, Xe, Ar, He, and Ne. For field and partial chemical analyses procedures, water samples are taken at the wellhead; for complete chemical and gas analysis, samples are taken from the bedding interval of the water-bearing stratum with the aid of deep sampler units, such as the PD-03 or PD-3M.

Fissure-Vein Water-Pressure Systems

The determining factor in GWF fissure-vein hydrodynamics is fracture tectonics (the system of deep fractures and smaller ones that feather out from them). Areas where fractures develop are usually characterized by positive temperature anomalies and greater thermal water inflow into the wells. Due to the complex geological structure and hydrodynamics of fissure-vein GWFs, the exploitable reserves of geothermal water at such fields are determined on the basis of empiric evaluations.

The following factors are studied at each fissure-vein field: (1) size of the field, area, and depth (zone of maximal inflows into the wells); (2) geologic structure of the field (bedding of rock and the petrographic composition, stratigraphic age of the rock, spatial location of large faults, type of faulting, etc.); (3) geothermic conditions; (4) magnitude of the stable flow rate of geothermal water or the enthalpy of steam-water mixture during simultaneous testing of wells drilled at the field; and (5) analysis of the geothermal water's chemical composition and dissolved gases, its corrosive properties, and salt deposition (geyserite).

In the USSR, exploration data have been accumulated for thermal and superheated water in fissure-vein type water-pressure systems in regions

67

of modern volcanic activity* (Kamchatka, Kurile Islands). As in the case of stratal systems, the tapping of water-bearing intervals in fissure-vein systems when drilling exploratory wells is carried out primarily on the basis of observations made during the drilling process and field-geophysical investigation. The specific aspects of exploring fissure-vein systems that contain superheated water are influenced by the fact that well production at such fields is in the form of steam-water mixture (in stratal conditions, water is in a liquid phase because formation pressure substantially exceeds the elasticity of saturated steam). The methodology for testing steam-water flowing wells differs from those techniques applied during flow tests of wells whose output consists of a monophase fluid.

After completion, the well must be in a state of "rest" (valve shutoff) for an extended period of time to measure the steady-state temperatures in the well and the initial formation pressure. Well tests are carried out in two stages:

(1) First, the steam's pressure range at the wellhead is determined and the behavior of the steam-water mixture output at different pressures is adjusted;

(2) Second, the optimal output of the well, the calorific value of the steam-water mixture and its chemical composition are determined.

When output of a steam-water producing well is initiated, the water rises along the well stem to a level where it boils, and a steam-water mixture is formed inside the stem. As this mixture travels toward the wellhead the quantity of steam increases, with a simultaneous increase in the mobility of the mixture, which can sometimes travel upward at a

---

*Fissure-vein GWFs in alpine fold areas (the Caucasus, Pamir, Tien-Shan, etc.) have relatively small reserves of low temperature that are used for balneologic purposes.

rate of several hundred meters per second. When calculations are carried out, the boiling process of superheated water and the flow of steam-water mixture is viewed as adiabatic boiling. This is because when heat is carried out at a rate of thousands or even tens of thousands Kcal/s, heat losses do not exceed 1-10 Kcal/s [12]. For this reason, measurement of the steam-water mixture enthalpy during the well test is practically equal to enthalpy of formation water (specific calorific value of water in a temperature range from 0 to 250°C is approximately 1 cal/g · °C). The amount of steam in the steam-water mixture is determined by the magnitude of enthalpy and pressure, and is calculated according to formulas used in thermotechnics. When the steam content is known, it is possible to calculate the specific weight, the specific volume, and the velocity of the steam-water mixture—if the discharge outlet flow rate and profile are known. The calculation of steam-water mixture (swm) parameters is carried out according to for-mulas II.17-II.20 [12]:

$$i_{swm} = i'(1-x) + i''x \quad ; \qquad (II.17)$$

$$\gamma_{swm} = \frac{1}{\dfrac{1-x}{\gamma'} + \dfrac{x}{\gamma''}} \quad ; \qquad (II.18)$$

$$v_{swm} = \frac{1}{\gamma_{swm}} = \frac{1-x}{\gamma'} + \frac{x}{\gamma''} = (1-x)v' + xv'' \quad ; \qquad (II.19)$$

$$W_{swm} = \frac{Qv_{swm}}{S} \quad ; \qquad (II.20)$$

where i_{swm} is the enthalpy of the steam-water mixture, i' = water enthalpy, i" = steam enthalpy, Kcal/kg; x = steam content in 1 kg of steam-water mixture (in kilogram parts); γ = density of the steam-water mixture; γ' = water density; γ" = steam density, kg/m^3; v = the specific volume of steam-water mixture, v' = specific volume of water, v" = that of steam, m^3/kg; W = the velocity of the steam-water mixture, m/s; Q = flow rate of the steam-water mixture, kg/s; S = the circular cross-section area, m^2. The values i', i", γ', γ", v', v" are determined using tables for dry saturated steam, in accordance with pressure and temperature.

Because temperature and pressure are mutually determining parameters in systems with saturated steam (each value of pressure corresponds to a strictly determined temperature level), one of the parameters (pressure or temperature) is measured at steam-water wells, and the other is ascertained using thermotechnical tables. The pressure at the wellhead of a flowing steam-water well corresponds to the pressure of saturated steam and is determined through thermal parameters and the flow of the steam-water mixture, and also by the conditions under which the mixture is discharged into the atmosphere. Thus, the minimal pressure at the wellhead (practically equal to atmospheric pressure) is achieved during free vertical spouting; when the diameter of the outlet is decreased, the lower pressure limit is increased. The maximal value of the steam-water mixture pressure is equal to the calculated value, which corresponds to the temperature of the thermal water that has been penetrated by the well. The operating pressure at the wellhead of steam-water wells is always significantly below this value: for example, at the maximal temperature registered at Well 14 at the Pauzhetka GWF, 195°C, the saturated steam pressure corresponds to

70

14.2 atm; at the same time the measured working pressure at the wellhead of this same well is only 6.75 atm. Smaller values of working pressure in contrast to maximal calculated pressures are explained by the fact that the point when steam is formed occurs at a significant depth, and as the steam-water mixture moves, part of the internal steam energy is transformed into kinetic energy; in addition, part of the steam-water mixture energy is lost as a result of hydraulic resistance. Overall, the depth of steam formation increases with the increase of the temperature of the superheated water and with the increase in well diameter; it is decreased with the increase of hydraulic pressure. Based on test data from the Pauzhetka field, steam formation in operating wells begins at a depth ranging from tens to hundreds of meters. Test results of steam-water wells are shown in the diagram that characterizes the empiric relationship between the steam-water flow rate and the steam pressure at the wellhead (figure 7). The principles discussed above pertain to steam-water wells with a stable output.*

The principal problem occurring when the flow rate of a steam-water well is being determined is that direct measurements of the volume of steam-water mixture in the flow do not give satisfactory results. For this reason, steam and water are separated for measurement purposes, or conversely, the steam-water mixture is condensed into a monophase media.

Hydrochemical exploration of steam-water wells is also tied to the specific features of steam-water well dynamics: the formation of a two-phase output determines the differentiation of the chemical composition of superheated water tapped by the well. This is because low molecular

_____

*There are also wells with pulsing output, and those with irregular behavior similar to the patterns in natural geysers.

71

FIGURE 7

OUTPUT OF STEAM-WATER WELLS AT THE PAUZHETKA FIELD
(WITH OUTPUT RELATIONSHIP TO PRESSURE AT WELLHEAD)

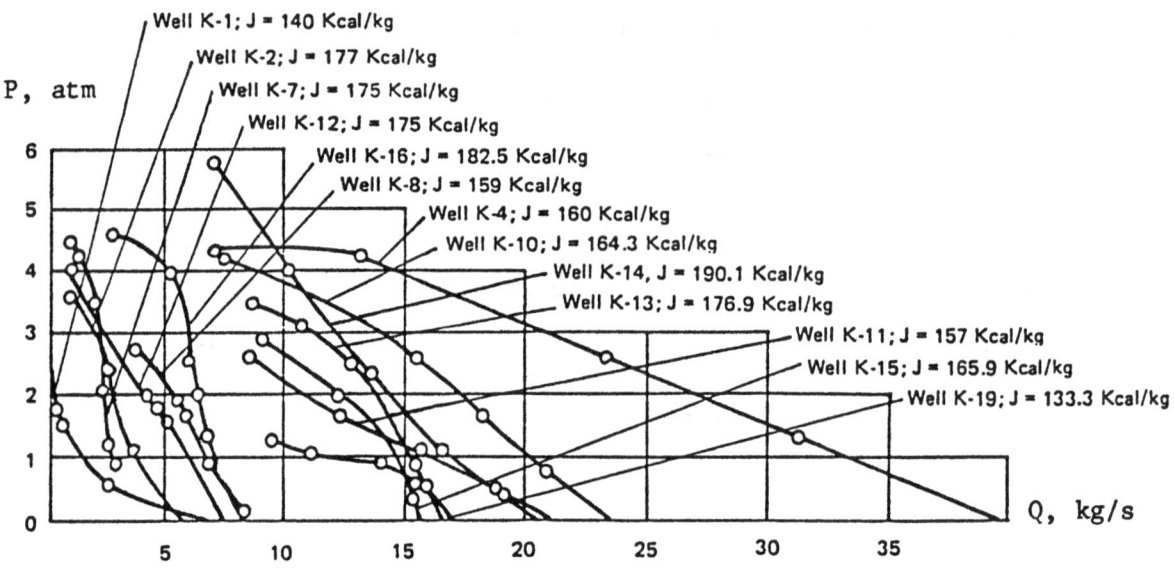

(J = enthalpy)

Source: Frolov, N. M., et al. Metodicheskiye ukazaniya po izucheniyu
termal'nykh vod v skvazhinakh. Moscow, Nedra, 1964, p. 31.

substances dissolved in the water are transformed into the steam phase. Water becomes degasified at the same time and in addition, loses some salts; with respect to the basic mass of salts, however, the water becomes more concentrated.

In fissure-vein water-pressure systems, thermal water in regions of recent and contemporary volcanic activity has a composition ranging from sodium sulfate-hydrocarbonate to sodium chloride with carbon dioxide, hydrogen-sulfide or nitrogen-methane dissolved gases. Mineralization of nitric thermal springs in montane fold structures is usually low, 1-2 g/l. Thermal and superheated water in fissure-vein systems and water in stratal water-pressure systems are both viewed as multi-purpose resources, and for this reason they are subjected to extensive chemical analysis. This analysis includes the same elements tested in thermal water of stratal systems (see preceding section).

5. Assessment of Commercial Thermal Water Reserves

The quantitative evaluation of a GWF, necessary for composing a plan for using the thermal water as a source of thermal or kinetic (superheated water) energy, is carried out via hydrodynamic calculations, simulation, or empirical evaluations. The first two methods are employed to evaluate the exploitable reserves of thermal water in stratal systems, and the empirical method is used to evaluate fissure-vein GWFs.

Degree of depth of study of the GWF is expressed by categories of exploitable reserves. Potential and hypothetical exploitable reserves of thermal water (categories C_2 and C_1) are determined by using approximated hydrothermic parameter evaluations, which are usually obtained by an analogy method. Industrial-level commercial reserves of thermal water in A and B categories are usually evaluated only on the basis of parameters established following study of the producing formation.

Hydrodynamic Estimation Methods

Estimation of exploitable thermal water reserves in strata systems is drawn from the same theoretical bases as the estimation of cold water reserves: filtration is viewed as being monophasal and isothermic. Consequently, methodology with a foundation in elastic filtration regime (C. V. Theis, V. N. Shchelkachev) is used for evaluation of GWF reserves as well. When estimations of exploitable thermal water reserves are made using the hydrodynamic calculation method, analytical mathematical formulas derived from the solution of differential equations that express water budget under varying hydrogeologic conditions are used. To solve the problem concerning inflow of subsurface water to the well in strata water pressure systems, the Theis, or basic elastic infiltration regime theory formula, is used:

$$\Delta P = P_c - P_{(r,t)} = \frac{Q\mu}{4\pi kh}\left[-E_i\left(\frac{r^2}{4at}\right)\right] \quad , \qquad (II.21)$$

where Q = the constant well production, cm^3/s; P_c = the initial formation pressure, kg/cm^2; $P_{(r,t)}$ = current pressure at any point in the stratum, located from the axis of the well at a distance of r, cm, at point in time t, sec, which are measured from the moment the well starts to operate; k = permeability, darcy; h = effective thickness of the water-bearing formation, cm; a = piezoconductivity index of stratum, cm^2/s; E_i = integral exponential function symbol, the value of which is shown in mathematical manuals.

In hydrodynamic calculation practice, the value of E_i argument function is the dimensionless hydraulic resistance R, usually less than 0.1; consequently this function is approximated with a high degree of accuracy by the logarithmic relationship

$$-E_i\left(-\frac{r^2}{4at}\right) \approx \ln \frac{2.25at}{r^2} \qquad . \qquad (II.22)$$

Formula II.21 describes the radial liquid inflow to the well in the conditions of a uniform "infinite" stratum. For other forms of layers (semi-infinite, circular, quadrant, band, right angle, wedge), the dimensionless hydraulic resistance R is determined by "mirror reflection," image-well boundary techniques [33, 34]. In the USSR, the elastic regimen filtration theory and Theis equation have been widely used to solve applied hydrogeologic problems since the 1960s [33, 34, 37, 38].

When making evaluations of exploitable thermal water reserves, often calculations for anisotropic strata must be made. In a number of published works, various anisotropic strata are examined. Thus, there is a solution for an infinite anisotropic stratum with one rectilinear separation boundary [35] and for an infinite stratum with a circular separation boundary [36]. Methods for estimating different systems of interfering wells (linear row, circular, and area well systems) have also been developed [33, 34].

As a rule, with the use of hydrodynamic formulas, calculations of exploitable thermal water reserves are estimated for a 25-year period. Here the limiting factor in estimating the commercial reserves for the researched fields is the flow that can be produced by gushing: pump exploitation of geothermal wells is currently economically unfeasible.

Commercial reserves of thermal water in the Makhachkala, Ternair, and Groznyy GWFs were estimated by analytical calculations using hydrodynamic formulas derived from the equations of Theis. In addition, the sum productivity of the areal system of interfering geothermal wells (exploited for 25 years) was established. However, when potential geothermal reserves are evaluated in various regions, the limiting factors may be other criteria, e.g., when mapping potential exploitable reserves on the territory of the USSR, calculated drawdown of wells was taken to be equal to 100 m below the surface [16].

At the end of the 1960s and early 1970s, the VSEGINGEO Institute carried out a regional evaluation of hypothetical geothermal resources in the USSR [16]. Modified formulas derived from the Theis equation were utilized to make hydrodynamic estimates of the reserves in stratal water pressure systems. In addition, groups of arbitrarily selected wells (thermal-water intakes) were represented as generalized systems. In many instances, the solution for production from a radial system of wells in an infinite stratum was used.

Estimates of overall hypothetical thermal water reserves for some USSR regions (Western Siberia, Sakhalin, and the Fergana valley) were also carried out on the basis of the so-called "barrel" method [38]. Here a hypothetical schematic was used, in which water-bearing structures appeared to be outlined by impermeable contours, and where numerous water intake areas were uniformly distributed over the entire area (within a square network). In addition, each of the structures was viewed as being hydrodynamically whole. This method made it possible to determine the extreme value of hypothetical reserves. This value characterized the overall potential of the water-bearing structure. The calculation formula for the given method may have the following form:

$$Q = 10^6 \frac{k'mSF}{at} \quad , \quad \text{(II.23)}$$

where Q = hypothetical exploitable reserves, m^3/day; F = area, m^2; S = the designated drawdown toward the end of the calculated exploitation period, taken to be equal to 100 m below the surface; t = the calculated period of exploitation, taken to be equal to 10^4 days; $k'm$ = the hydroconductivity factor of the structure, m^2/day*; a = piezoconductivity index, m^2/day (in case of absence of data, taken to be 10^5).

Hypothetical thermal water reserves in small areas that have been calculated using the "barrel" method are usually too low, since natural resources of the system (external inflows) are not taken into account. Stratal-type GWFs are subdivided according to water recharge rates into low flow (less than 50 l/s), average flow (50-100 l/s), and high flow (above 100 l/s). Productivity estimates of water recharge areas were carried out using the standard 5-well network distributed over a 25 km² area. The reduced radius of a selected standard water-intake was 405 m. More frequently productivity estimates for the water recharge area were made using the infinite stratum formula II.21. Flow rates of standard water intakes, calculated for different regions, were shown on layer

---

*The following relationship exists between the permeability and the infiltration index:

$$k'[m/day] = 0.864 \, k[darcy] \frac{\gamma[g/cm^3]}{\mu[cp]} \quad . \quad \text{(II.24)}$$

Taking formula II.24 into account, in order to switch from the hydroconductivity factor in darcy·cm/cp to the m^2/day unit, the following relationship may be used:

$$k'm[m^2/day] \approx 0.01 \frac{k[darcy] \, h[cm] \, \gamma[g/cm^3]}{\mu[cp]} \quad . \quad \text{(II.25)}$$

maps and zonation maps for hypothetical thermal water reserves.

In Eastern Precaucasia, many GWFs are located comparatively close to thermal water outcrops, where cold water infiltrates into them (the Makhachkala, Ternair, Izberbash, and Groznyy GWFs, which are limited to Middle Miocene deposits, are found at a distance of 10-30 km from the recharge points). Evaluation of the period during which isotherm with a lower value will draw near a thermal water intake point was carried out using the H. A. Louwerier method [26]. Here it was shown that even with rather substantive mobility rates of water along the stratum (about 90 m/year), decrease in temperature of the exploited sites found within 10 km of the recharge zones may occur 300 years after production is initiated. In addition, this drop will be approximately 9°C, and when thermal water intakes that are more than 10 km away from the recharge zones are exploited over a period of 50-150 years, there will be practically no drop in temperature.

Simulation Method

Mathematical filtration simulation methods for subsurface water make it possible to estimate the natural hydrogeologic conditions of research sites more accurately; these conditions include water and physical characteristics of the filtration media, the compound nature of the boundary conditions (geometry and hydraulic features), and the effect of artificial hydraulic factors. The application of simulation methods is most effective in solving hydrogeologic problems relating to the evaluation of well productivity, in cases of anizotropic filtration in water-bearing horizons, and interaction of different water-bearing horizons through semi-permeable water confining strata. Simulation also facilitates estimation of large well systems in which well production is started at different times. When there is adequate information present concerning the hydrogeothermic conditions in a GWF, the simulation method may be utilized to evaluate commercial reserves.

Network electric models are usually used to solve filtration problems tied to the evaluation of exploitable reserves of subsurface (thermal) water. Definition of the parameters by simulation and determination of the schematic of rational well distribution requires repeated examination of electric model elements, which is effectively accomplished with the aid of network-type models. Typically resistance and capacitive networks (RC networks) are utilized.

An example of evaluating thermal water reserves with the simulation method was the estimation of hypothetical thermal water reserves of the Rionskaya depression in Soviet Georgia (Transcaucasia) [39]. When filtration processes were simulated for the Neocomian water-bearing structure, which is a most promising site, the hydrogeological role of faults was evaluated and the magnitude of discharge and recharge of the water-bearing complex was defined. The following equation was derived for the model:

$$\frac{\partial}{\partial x}\left[T(x,y)\frac{\partial H_{rh}}{\partial x}\right] + \frac{\partial}{\partial y}\left[T(x,y)\frac{\partial H_{rh}}{\partial y}\right] = 0 \quad , \qquad (\text{II.26})$$

where T = the hydroconductivity factor of the water-bearing structure; H_{rh} = the value of the reduced head for the selected comparison plane.

The boundary conditions are shown in figure 8. The network spacing changed from 1,500 to 6,000 m, and was decreased in regions with high density of well siting. The resistance networks were calculated on the basis of the hydroconductivity factor map, which was corrected during

79

FIGURE 8

SCHEMATIC MAP OF REDUCED PRESSURES
OF THE NEOCOMIAN WATER-BEARING FORMATION (RIONSKAYA DEPRESSION)

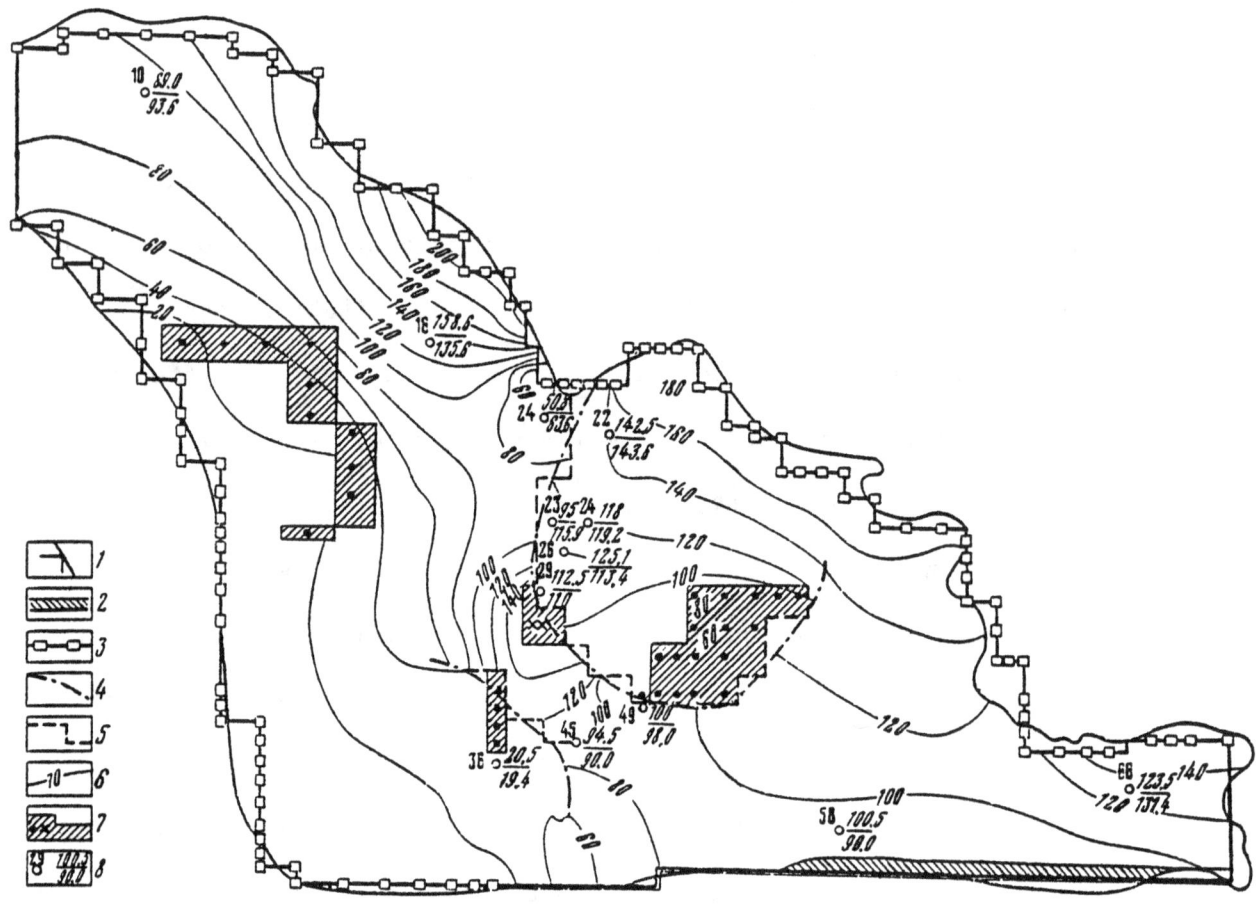

1   Outcrop contour with the condition: $H|_{g_1} = H(x,y)$

2   Discharge zone of the Neocomian structure along the
     North-Adzharskiy fault with boundary conditions: $Q|_{g_2} = Q(x,y)$

3   External boundary of modeled section (along the bottom of the Black
     Sea) with boundary conditions: $H|_{g_3} = 0$

4   Fault contour that causes a break in the continuity of the filtra-
     tion flux

5   Inner contour of the modeled fault

6   Isolines of modeled reduced fault

7   Recharge zones and discharges derived as a result of the modeling

8   Wells (on the left, number; on the right, value of reduced head, in
     m; numerator, factual; denominator, simulated

Source: Krashin, I. I. and D. I. Peresun'ko. Otsenka ekspluatatsionnykh
          zapasov podzemnykh vod metodom modelirovaniya. Moscow: Nedra,
          1976, pp. 178-9.

80

modeling of discharges. The model also represented the contours of faults disturbing the continuity of the filtration flux.

Having set additional recharge and discharge, where this was possible, the problem was solved by deriving the coincidence of modeled and calculated values of reduced heads for the wells. The representation of the natural regime of subsurface waters of the Neocomian water-bearing formation made it possible to understand its hydrodynamic conditions--the nature of thermal water flow, the role of faults, and the magnitude of recharge and discharge.

In order to evaluate the exploitable reserves of thermal water, modeling was applied for two types of production methods: gusher and forced pumping. For predicted variants, the modeling period comprised 45 years: from 1953 to 1998. The solutions of forecasting problems consisted in deriving the maximum possible water production using different well exploitation methods. In this instance, the evaluation of exploitable resources was conducted for water intakes (standard water intakes, 25 km^2 in area, located on the site of operating wells and at projected sites) with the use of pumping, causing a drawdown 100 m below the land surface. These calculations showed that the flow of the exploited wells is assured as a result of elastic reserves of the Neocomian water-bearing complex and by drawing upon additional reserves of subsurface water created by reduction of water loss in the discharge complex (figure 9). Table 6 shows results of balance calculation for one of the variants calculating the reserves under forced pumping exploitation of standard water intakes, with a drawdown to 100 m below the land surface.

At the end of the 1960s, SevKavNIIgaz (Stavropol) performed evaluations of hypothetical exploitable reserves of thermal water in the

PROJECTED (1997) DROP IN LEVEL OF SUBSURFACE WATER AT THE NEOCOMIAN
WATER-BEARING FORMATION WITH PUMPED EXPLOITATION RESULTING IN A DRAWDOWN
OF 100 m BELOW THE LAND SURFACE

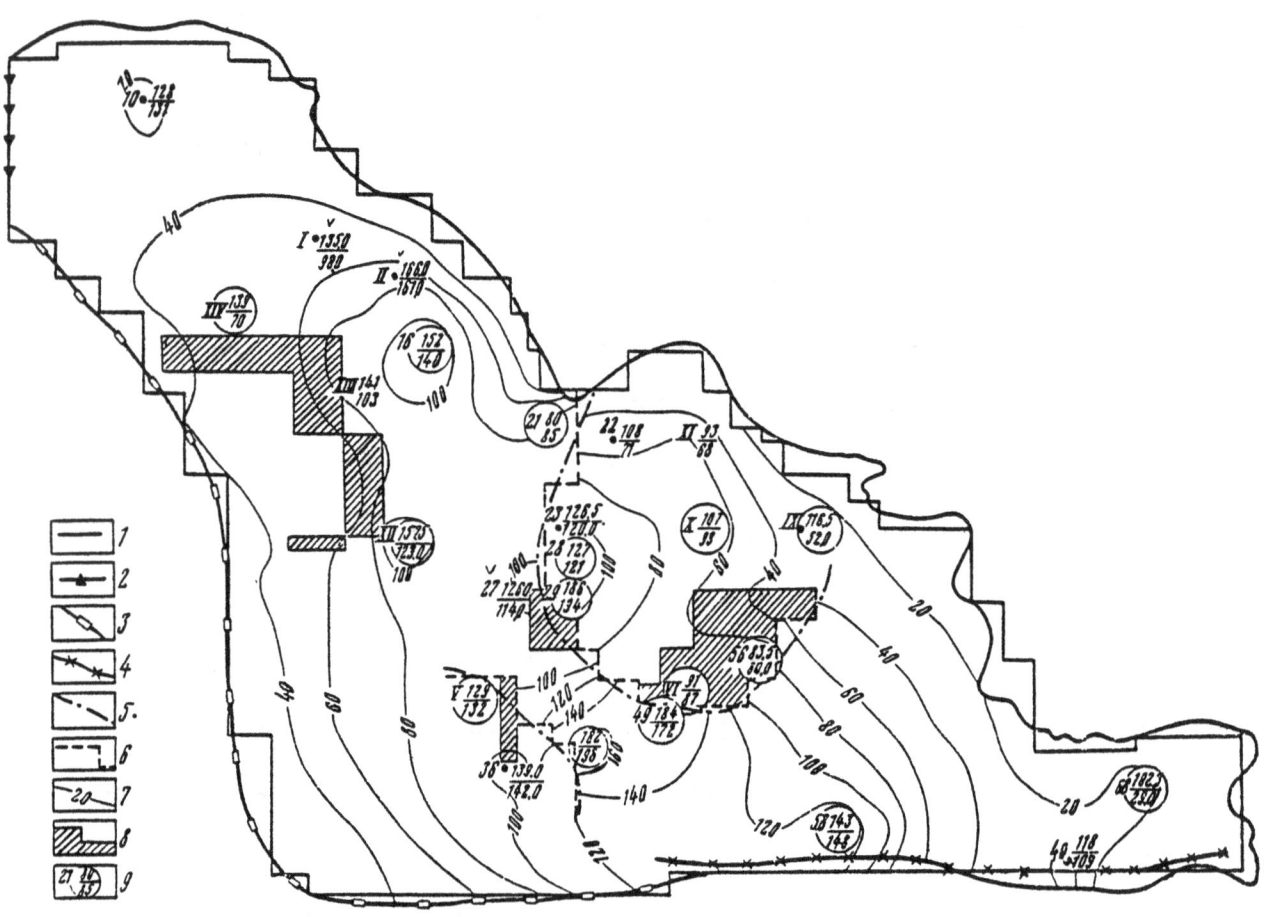

1. Distribution contour for the water-bearing structure with boundary
    conditions: Q_{g_1}, $S < S_0 = \dfrac{S_{g_1}(x, y)}{F_g(x, y)}$; O_g, $S > S_0 = $ const

2. External boundary of the modeled region with boundary conditions:
    $Q_{g_2} = 0$

3. External boundary of the modeled region with boundary conditions as
    in contour g_1

4. Fault at external boundary with boundary conditions as in contour g_1

5. Actual fault contour that creates the break in filtration
    continuity

6. Modeled contour of fault

7. Isolines of drops in underground water levels for 1997

8. Recharge and discharge zones with boundary conditions as in contour
    g_1

9. Recharge area (in circle) and well (left - number; right: in the
    numerator - allowable drop, in the denominator - modeled level)

Source: Krashin, I. I. and D. I. Peresun'ko. Otsenka ekspluatatsionnykh
         zapasov podzemnykh vod metodom modelirovaniya. Moscow: Nedra,
         1976, pp. 182-83.

TABLE 6

BALANCE CALCULATION WHEN PIEZOMETRIC LEVEL DROPS TO 100 m

Balance Indicators	Impulse, %						
	23	30	39	41	45	50	89
	1964	1968	1972	1973	1975	1978	1998
Water Production							
l/s	130	330	330	2,000	2,000	2,000	2,000
Recharge							
Percent	28.5	22.6	62.3	28.5	54.9	68.9	94.5
l/s	37.5	74.7	205.6	569.4	1,099.4	1,397	1,890
Capacitive Reserves							
Percent	70	--	45.1	75.2	45.1	30.2	5.5
l/s	91	--	149	1,504	902	603	110

Karaganian water-bearing formation in Eastern Precaucasia using a numerical method and a Minsk-22 computer (S. S. Gatsulayev, S. L. Polevoy, et al.).

Empirical Evaluation Method

The flow of thermal water in fissure-vein water-pressure systems occurs in turbulent or transient (laminary turbulent) flow patterns. In this instance, filtration theory equations based on linear Darcy law are inapplicable. Great difficulties also arise when determining the volume and shape of the fissure system complex and the cavities along which the thermal water circulates. In this case, empirical methods acquire a great importance when evaluations of thermal water reserves in fissure-vein systems are made. As research of fissure-vein GWFs shows, exploitable thermal water reserves in such systems exceed their natural discharge by several times. For example, data analysis for over 50 exploratory fissure-vein fields showed that the sum flow of thermal water produced by the wells is higher than natural discharge by a factor of 3 to 30 and greater [16].

The evaluation of hypothetical thermal water reserves in fissure-vein formations is carried out on the basis of the natural discharge; in addition, the flow increase coefficient as a result of drilling is taken into account, and is taken to be from 2 to 3, depending on the nature of the discharge center and the accompanying signs of thermal water and vapor-hydrothermae discharge. Industrial evaluation of commercial reserves in fissure-vein GWFs is carried out on the basis of factual data for well productivity; these are established as a result of lengthy synchronous flow tests of steam-water mixture and thermal water. Such tests are usually carried out over a period of one year in order to establish the influence of meteorologic and hydrologic factors on the productivity of the wells, to study the influence of the wells on each other, and to determine the optimal production at which pressure in the system will be stabilized. The empirical method was used when industrial reserves were evaluated at the Paratunka thermal site and at the Pauzhetka superheated water field [42].

Hydrogeothermic Parameters Used to Evaluate Exploitable Geothermal Reserves

An entire set of parameters that characterizes the natural hydrodynamic and thermal-physical properties of the water and the reservoirs must be defined in order to evaluate the exploitable thermal water reserves. These parameters may be defined separately or in the form of complexes (table 7); methods of determining these parameters are outlined below.

Hydrodynamic Methods

Hydrodynamic methods, which are based on data for thermal water flow tests, make it possible to determine the mean values of hydroconductivity and piezoconductivity of geothermal strata. Parameters are determined through the solution of inverse hydrodynamic problems. In addition, the graphic-analytical method is widely used to solve equation II.21; this is known as a pressure build-up (drawdown) test.

Another basic method for determining hydrodynamic parameters of stratal systems is the multiple-well transient test, which consists of recording pressure changes over time in observation wells while simultaneously recording the flow rate of the active well over time. This test also makes it possible to determine the piezoconductivity index, a. Soviet researchers have derived numerous transient methods for determining hydrodynamic parameters of strata [S. N. Nazarov, I. A. Charnyy, G. I. Barenblatt, V. P. Borisov, S. N. Businov, et al.].

85

TABLE 7

BASIC HYDROGEOTHERMIC PARAMETERS OF GWFs

Water-Bearing Rock		Thermal Water	Complex Features	
Stratal and Fissure-Vein Systems	Stratal Systems	Stratal and Fissure-Vein Systems	Stratal and Fissure-Vein Systems	Stratal Systems
Effective thickness, h	Permeability, k	Enthalpy, i	Formation temperature, t	Filtration index, k'
Specific electrical resistance, ρ	Porosity	Volume weight, γ	Formation pressure, P_r	Filtration resistance, $\dfrac{\mu}{kh}$
Lithologic and petrographic composition	Volume weight, γ_n	Dynamic viscosity, μ	Specific well yield, q	Piezoconductivity index, a
	Elastic compression coefficient, β_r	Elastic compression coefficient, β_c	Heat conductivity coefficient, λ_t	Elastic capacitance coefficient, β^*
	Granulometric composition	Overall mineralization level and salt composition		
		Specific electrical resistance, ρ_w		
		Elasticity of dissolved gases, P_g		
		Gas saturation level, v_o		

Hydraulic Method

The hydraulic method, also known as the steady extraction method, is also widely applied to establish the hydrodynamic parameters of strata. During flow tests in wells, the well's yield and pressure become practically constant (quasi-steady) after a certain period of time. Under elastic filtration conditions at the onset of a quasi-stationary state of flow, the relationship between well production and pressure can be described by the Dupuit formula, and the hydroconductivity factor may be derived from the expression

$$\frac{kh}{\mu} = q \; \frac{\ln \frac{R}{r_{red}}}{2\pi} \; . \qquad (II.27)$$

Specific rate of flow, $q = \frac{Q}{\Delta p}$, is determined on the basis of geothermal borehole flow tests. In addition, the magnitude of depression of ΔP may be determined on the basis of drawdown at the wellhead, provided corrections are taken into account (using the methodology shown in part 4 of this chapter). The value R, radius of the cone of depression, is usually arbitrarily taken to be 10 km; r_{red}, reduced radius of the well, is determined according to diagrams of V. I. Shchurov, taking the imperfect nature of the well screen and the level to which the stratum has been exposed (skin effect) into account.

For technological calculations related to the development of GWFs it is necessary to determine the mean heat conductivity coefficient of the rock along the borehole. Laboratory examinations of core samples do not provide reliable information concerning the thermophysical properties of the cross-section that has been exposed by the well, which usually consists of frequently alternating layers with varying lithologic composition, density, and water saturation. For this reason, average thermal conductivity coefficient was determined over the entire stratal section or along individual intervals on the basis of data derived during flow tests in wells operating in quasi-steady thermal and hydraulic regimes [24]. When a quasi-steady flow of liquid sets in, the heat distribution process in a geothermal well is viewed as being similar to the distribution of heat in a cylindrical wall. Heat transfer in a producing well occurs primarily in a radial direction via convectional heat exchange between the thermal water and the wall of the well, and as thermal conduction from the wall of the well to the surrounding rock. Assuming that temperature distribution along a borehole in a natural geothermic field is approximately linear, and that the amount of heat lost by thermal water* as it moves from bottomhole to the wellhead is equal to the amount of heat that has been given off from its lateral surface to the rock surrounding the bore, the thermal balance in the operating well is characterized by the equation

$$\int_0^H \frac{2\pi \lambda_t (t_{wh} - t_{nl})}{\ln \frac{R_t}{r_o}} \, dx = \int_{t_{wh}}^{t_f} WC\gamma \, d(\delta t_1) \quad , \qquad (\text{II}.28)$$

from which the mean heat conductivity coefficient for the total rock exposure is equal to:

---

*The amount of heat lost as a result of the transformation of hydraulic energy into thermal energy, as well as heat losses through the external part of the casing, are insignificant.

88

$$\lambda_{t(ave)} = \frac{WC\gamma(t_f - t_{wh}) \ln \frac{R_t}{r_0}}{2\pi H(t_{wh} - t_{nl})} \quad . \tag{II.29}$$

The formula for determination of the heat conductivity coefficient for individual strata exposed by the well has been derived in an identical fashion:

$$\lambda_{t(1-2)} = \frac{WC\gamma(t_{c_1} - t_{c_2}) \ln \frac{R_t}{r_0}}{2\pi\Delta H(t_{wh} - t_{nl})} \quad . \tag{II.30}$$

In formulas II.28-30, the following designations have been used: $\lambda_{t(ave)}$ = the average heat conductivity coefficient for the geological section; t_f = formation temperature; t_{wh} = temperature at the wellhead of a flowing well; t_{nl} = temperature at the neutral (constant temperature) layer level; W = the flow rate of liquid in a unit of time; C = thermal capacity of the fluid; γ = the specific liquid weight; δt_1 = changes in flow temperature; R_t = the radius of thermal influence; r_0 = well radius; dx = differential of flow length in the well.

The value of the entire temperature pressure does not practically depend on the nature of changes in the temperature of rock under natural conditions in the interval between the neutral layer and the bottomhole. This makes it possible to determine the magnitude of the mean heat conductivity coefficient of the cross-section, without taking into account temperature distribution along the bore of geothermal wells that have undergone a "rest period." All values included in the right side of equations II.29 and II.30, with the exception of R_t, are determined on the basis of geothermal water flow tests. The radius of the well's thermal influence is evaluated using the formula

$$R_t = 2\sqrt{\alpha_t \tau} \quad , \qquad\qquad (II.31)$$

where α_t = the thermal diffusivity coefficient of rock, m^2/hr; τ = the operating time of flowing well, hr.

As calculations and experimental data show, value R_t may be viewed as being equal to 2-3 m with substantial accuracy [41]. Based on the data derived during study of geothermal wells, mean heat conductivity values have been calculated using formula II.29 for some of the sand-clay sedimentary sections (table 8). Heat conductivity coefficient values derived on the basis of flow tests during a steady hydraulic state match well with data from laboratory efforts.

Geophysical Methods

Geophysical methods of determining hydrodynamic parameters of water-bearing strata are indirect methods. Such strata parameters as effective thickness, porosity, permeability, formation temperature, specific resistance of formation water in the well, etc., may be resolved with the aid of geophysical techniques. The effective thickness of exposed reservoirs is determined on the basis of a series of geophysical and geological tests. Sand reservoirs are determined on the basis of SP, GRL curves, caliper logging, and by AR data, GGRL, and acoustic logging. In carbonate rock, the use of geophysical methods alone to determine effective thickness is inadequate. In this case, a series of special methods are used, including photo-logging, gas logging, study of core samples and use of downhole flowmeter.

Porosity is determined using GGRL and NGRL diagrams, as well as the acoustic logging data resistance method. Empiric relationships are

TABLE 8

CALCULATION OF AVERAGED ROCK THERMAL CONDUCTIVITY COEFFICIENTS USING FLOWING THERMAL WATER TEST DATA
(BASED ON FORMULA II.29)

Flow Rate		Formation Temperature $(t_f, °C)$	Water Temperature at Wellhead $(t_{wh}, °C)$	Depth of Well (H, m)	$\Delta t=(3)-(4)$, (°C)	$Q=10^3 \cdot (2) \cdot (6)$ (Kcal/hr)	Full Temperature Head $\Delta t_H=(4)-12$ (°C)	$6.28 \cdot (5) \cdot (8)$ $(m \cdot °C)$	$\lambda_{t(ave)}=3.44 \cdot (7):(9)$ (Kcal/m·hr·°C)	$\lambda_{t(ave)}$ on Area (Kcal/m·hr·°C)
m^3/day	m^3/hr									
1	2	3	4	5	6	7	8	9	10	11
		$C = 1$ Kcal/kg·°C; $\lambda=10^3$ kg/m³; $C\gamma = 10^3$ Kcal/m³·°C; $t_{n1} = 12°C$; $R = 3$ m; $r_0 = 0.1$ m; $\ln \frac{R}{r_0} = 3.44$								
						Makhachkala Area				
						Well 136				
311	12.96	76.5	64	1408	12.5	$1.62 \cdot 10^5$	52	$4.6 \cdot 10^5$	1.22	1.32
						Well 53				
810	33.75	76	70	1460	6	$2.02 \cdot 10^5$	58	$5.3 \cdot 10^5$	1.31	--
						Well 43				
655	27-29	74.2	67	1375	7.2	$1.96 \cdot 10^5$	55	$4.75 \cdot 10^5$	1.42	--
1320	35	74.2	69	1375	5.2	$1.82 \cdot 10^5$	57	$4.92 \cdot 10^5$	1.35	--
						Ternair Area				
						Well 22				
1233.8	51.41	69	65	1517	4	$2.06 \cdot 10^5$	51	$4.86 \cdot 10^5$	1.46	1.50
						Well 12				
1489.5	62.06	64	61	1380	3	$1.86 \cdot 10^5$	49	$4.25 \cdot 10^5$	1.51	--
						Well 95				
414	17.25	62	54	1181	8	$1.38 \cdot 10^5$	42	$3.115 \cdot 10^5$	1.52	--
						Izberbash Area				
						Well 15-T				
554	23.08	63.9	57	1390	6.9	$1.59 \cdot 10^5$	45	$3.93 \cdot 10^5$	1.39	1.41
951.3	39.64	63.9	59.5	1390	4.4	$1.74 \cdot 10^5$	47.5	$4.15 \cdot 10^5$	1.44	--
						Grozny Area (Khankal'skaya Valley)				
						Well 27/32				
498	20.75	98.2	93.1	850	5.1	$1.06 \cdot 10^5$	81.1	$4.33 \cdot 10^5$	0.82	0.88
1370	57.10	98.2	97	850	2.2	$1.26 \cdot 10^5$	85	$4.54 \cdot 10^5$	0.95	--
						Goytinskaya Area				
						Well 10-T				
376.7	15.7	84	70	1850	14	$2.16 \cdot 10^5$	58	$6.73 \cdot 10^5$	1.1	1.06
719.7	30	84	76.5	1850	7.5	$2.25 \cdot 10^5$	64.5	$7.5 \cdot 10^5$	1.03	--

NOTE: Calculations are shown on the basis of samples from thermal water-bearing strata in Middle Miocene deposits.

Source: [24].

broadly utilized to determine porosity; these are established for rock of specific type, age and area, based on geophysical data and core study.

The permeability coefficient is determined on the basis of the resistance method and the potential method of spontaneous and induced polarization. Due to the degree of their inaccuracy, permeability values determined by geophysical methods are used for approximate evaluations of this coefficient.

Laboratory Methods

Laboratory methods for determining aqueous and physical parameters are widely used in hydrogeologic research. Such parameters as porosity (open and dynamic), permeability, volume weight, granulometric composition, elastic compression coefficient, viscosity of stratal water, level of mineralization, chemical composition and gas saturation of water, and specific electrical resistance are also determined in laboratory conditions.

Modeling Method

In the modeling method for determining hydrogeologic parameters, the models are usually employed to solve inverse problems relating to natural and exploitation regime of subsurface (thermal) water. For this purpose, model parameters (electrical resistance and capacities) that correspond to filtration resistance and capacities of the media being modeled, as well as currents that correspond to the source values or discharge of the water-bearing horizon, are altered in such a way as to derive a process with the model (distribution levels or changes in them) that reflects what has actually been observed. This process requires repeated examination of the model's structural elements, using analog computer systems with an automated resistance network and a digital computer controlling unit.

CHAPTER III

HYDROGEOTHERMAL CONDITIONS IN REGIONS
WITH GEOTHERMAL POTENTIAL AND EXPLOITABLE RESOURCES

Hydrogeothermal research that has been carried out in the USSR over the last twenty years has made it possible to establish promising regions with geothermal water pressure systems, as well as to evaluate their overall exploitable reserves. The research effort that provides the principal generalization on the study of hydrogeothermal resources during this period is the estimation of hypothetical geothermal reserves that was done by the VSEGINGEO Institute Laboratory for Thermal Water [16]. Determinations of the geothermic gradient are shown on the USSR Geothermic Map, 1:500,000 scale, which was compiled by a research group from the AN SSSR Geologic Institute and the VSEGINGEO Institute, under leadership of F. A. Makarenko [43].

The distribution of thermal water development and geothermal exploration areas in the USSR is shown in figure 10; promising regions with geothermal water potential extend throughout the Soviet Union from its western borders (Carpathia) to Kamchatka and the Kurile Islands in the east (figure 11).

In stratal systems, thermal water lies in the deep sediments that fill depressions in crystalline basement, in foredeeps, and in intermontane basins, which in total area extend over more than one-third of the territory of the USSR. Of this area, 45 percent is located in Siberia and the Far East, approximately 35 percent in the European section of the Soviet Union, and 20 percent in Central Asia and Kazakhstan.

94

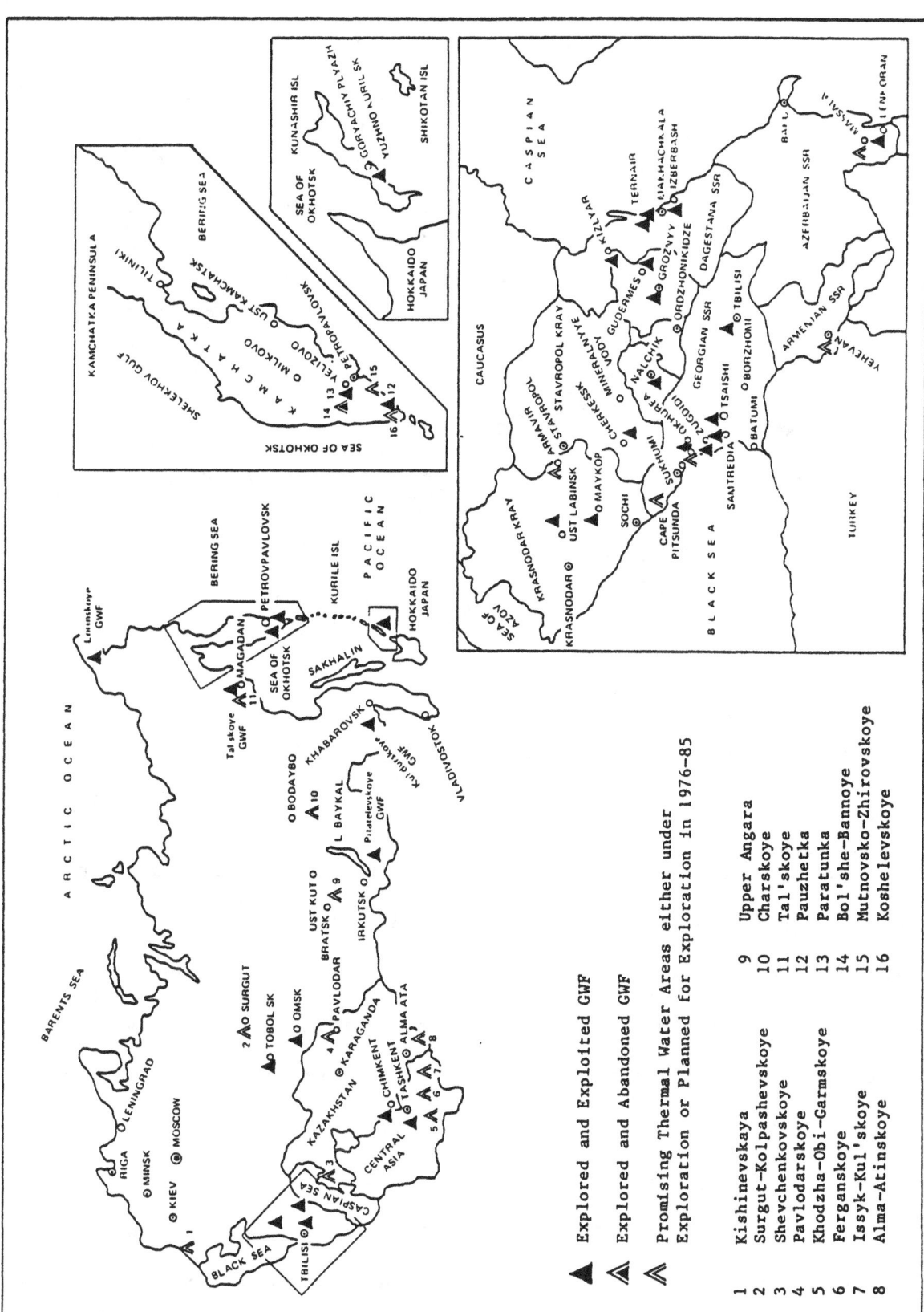

FIGURE 10

DISTRIBUTION OF GEOTHERMAL WATER FIELDS AND EXPLORATION AREAS IN THE USSR

95

FIGURE 11

DISTRIBUTION OF PROMISING THERMAL WATER AREAS IN THE USSR

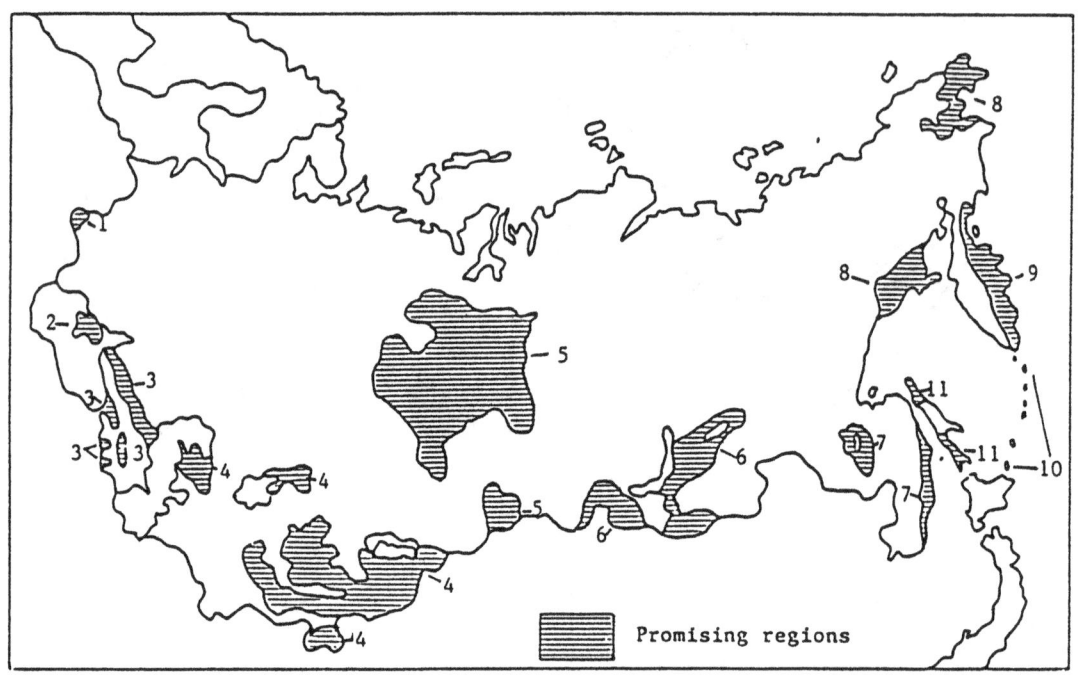

Promising regions

Note: Numbers correspond to chapter III subsections.

Source: Mavritskiy, B. F., et al. Resursy termal'nykh vod SSSR. Moscow, Nedra, 1975.

Fissure-vein thermal water is usually limited to metamorphic, volcanic-sedimentary, and igneous rocks of varying ages that form the fold regions and basements of artesian basins. This water is generally found in tectonic fault areas in small artesian systems. Vertical circulation dominates the dynamics of fissure and fissure-vein water, in contrast to the lateral circulation dominant in stratal systems. Fissure discharge is in the form of springs and sometimes steam jets, as well as hidden discharge on the basement surface beneath the artesian basins. In the USSR there are up to 150 groups of springs and individual flows with temperatures above 40°C, which are concentrated primarily in mountainous regions in the southern and eastern parts of the country and in the Transbaykal region. Thermal springs are also known in the northeast areas (Kolyma, Chukotka), where they penetrate to the surface through thick layers of permafrost (Salygan-Sylba, Lorinskiy, and Talskiy springs). The Kurile-Kamchatka volcanic zone is a unique site. Here, numerous hot springs and steam jets are tied to magmatic activity. The greatest spring flow volumes recorded in the USSR have been observed in this region and in the Sayan-Baykal fold area.

Geologic Settings of Geothermal Waters

Thermal waters in the USSR are distributed in 11 geologic provinces (from west to east): (1) Carpathia; (2) Crimea; (3) Caucasus; (4) Central Asia and Kazakhstan; (5) Western Siberia; (6) Southern part of East Siberia; (7) Transbaykal and the Amur Region; (8) the Northeast and Chukotka; (9) Kamchatka; (10) Kurile Islands; and (11) Sakhalin (figure 11).

1. Carpathia

The Carpathian area lies in the seismically active Alpine montane fold belt that extends from the Pyrenees and Alps in Western Europe to Tien Shan and the Pamirs in Central Asia.

A promising target for thermal water is the regionally distributed Sarmatsko-Levantinian water-bearing complex (Miocene) in the Chop-Mukachevskaya intermontane basin in Carpathia. Water-bearing strata include terrigenous and pyroclastic fragmental rocks (sandstones, siltstones, tuff, and tuffites). Thermal water production from wells reaches several hundred cubic meters per day, and water temperature is about 50°C. Overall mineralization of thermal water usually does not exceed 10-15 g/l.

In the Sarmat-Levantinian water-bearing complex, hypothetical exploitable reserves of thermal water have a temperature of 50°C and a flow rate of 545 l/s, and the thermal potential is estimated to be

$0.687 \cdot 10^{12}$ Kcal/year. Predicted regional water production in Carpathia is equal to the heat produced by burning 100,000 tons of conventional fuel annually, and capital outlays are expected to be returned in no more than 7 years [44].

2. Crimea

Promising thermal water-bearing structures of stratal-pore and stratal-fissure types are found on plain of the Crimea, which is a part of the Scythian (Skifskaya) plate. Sedimentary rocks from Jurassic to Quaternary age lie on basement of varying ages (Precambrian, Paleozoic, Early Mesozoic). With respect to the hydrogeothermic factors of Neocomian-Aptian, Danian-Paleocene and Eocene ages, water-bearing formations are of the greatest practical interest.

Water in Neocomian-Aptian deposits is confined to heterogranular sandstones and conglomerates that lie at depths of up to 2,500 m. Flowing thermal water rate, on the average, is 1,000 m^3/day, with water temperature to 70°C (Genicheskaya Well no. 5). Mineralization of water varies widely from fresh water to brines with a maximum mineralization of 80 g/l (Novo-Alekseyevskaya Well no. 6). These thermal waters usualy contained dissolved nitrogen and methane.

Thermal water in Danian-Paleocene formations is usually confined to fractured limestones, and more rarely, to marls. Well overflows vary greatly, from several m^3/day to 3,000 m^3/day. Water influx is reduced as the depth and the distance from the recharge area are increased. Subsurface formation temperatures may reach 55-70°C and higher, and the overall mineralization of thermal water varies from 1 g/l in eastern Crimea to 20-25 g/l on the Tarkhankutskiy Peninsula. Mineralized water is usually saturated with hydrocarbon gases.

In Eocene deposits, thermal water circulates in fractured lime-stones and marl, and near Pre-Sivash, is found in fine-grained, poorly cemented sandstone. Well production does not exceed 1,000 m^3/day. Mineralization in the recharge area is of the order of 1 g/l and increases with formation depth to 30 g/l. Overall hypothetical reserves of thermal water in the Crimea plain region is 2 m^3/s.

3. Caucasus

Structurally, the Caucasus region is subdivided into the Precaucasian platform, inside which the Azov-Kubanskiy and Tersko-Kumskiy basins are located; the folds of the Greater and Lesser Caucasus and Tamysh; and in Transcaucasia, the Rionskaya and Kurinskaya depressions.

Precaucasia

In the Precaucasian platform, the Azov-Kubanskiy and Tersko-Kumskiy artesian basins are separated by the Stavropol dome. Sedimentary thickness reaches 12-15 km. Both basins are of the piedmont type and have external pressure-forming recharge areas also in the piedmont areas in the Caucasus. Most promising water-bearing systems in the Precaucasia area are in Neogene deposits (Apsheronian, Pontian-Meotian, Karaganian, Chokrakian), and sediments of Lower Cretaceous age. The Aspheronian water-bearing formation is widely distributed in Eastern Precaucasia. Thousands of artesian wells 100-200 m deep supply farms. Water-bearing strata are usually loose sandstones, with gravel and pebbles. Water conductivity is of the order of 200 m^2/day. Over the entire area, water in Apsheronian deposits is usually fresh or slightly saline. Near the Caspian Sea (Kizlyar), Apsheronian formations are found at a depth of more than 1,000 m. Here, they are up to 700 m thick and contain slightly saline (to 2 g/l) thermal water (50°C).

The Pontian-Meotian water-bearing formation is widely found in the Azov-Kubanskiy and the Tersko-Kumskiy foredeeps and to the north of the Stavropol dome. Lithologics are sand, sandstone, and siltstone. The depth at which Pontian-Meotian formations are found does not exceed 2,500 m (in the Azov-Kubanskiy depression); water is fresh or slightly saline. The Azov-Kubanskiy foredeep is noted for the highest conductivity (200 m^2/day). Wellhead overflow rate is up to 1,000 m^3/day; water temperatures vary from 40 to 70°C. Hypothetical thermal water reserves are estimated to be 2.4 m^3/s.

The Karaganian water-bearing complex has major industrial significance in the southern and eastern parts of eastern Precaucasia (figure 12). Here the recharge area can be clearly traced along the northern slopes of the Caucasian range. Partial discharge occurs approximately 50 km to the south as thermal springs on the forerange. Water-bearing Karaganian sandstone formations become quite thick (up to several hundred meters) near the Caspian Sea, and have high permeability. These occur in a band approximately 50 km wide that extends along the piedmont of the Greater Caucasus.

The Karaganian water-bearing formation extends to a depth of 4 km, where water circulating in it is heated to a temperature exceeding 100°C. These deposits are dynamically flushed and contain fresh, brackish, and saline waters with maximum mineralization that is below 10-20 g/l.

The Chokrakian water-bearing formation, like the Karaganian, has commercial value as a source of thermal water in the foothills area of eastern Precaucasia. The recharge area is also located on the north slopes of the Greater Caucasus, where monoclinal Chokrakian rocks outcrop. At great depths, water in Chokrakian formations acquires a

FIGURE 12

HYDROGEOTHERMIC MAP OF THE KARAGANIAN WATER-BEARING FORMATION

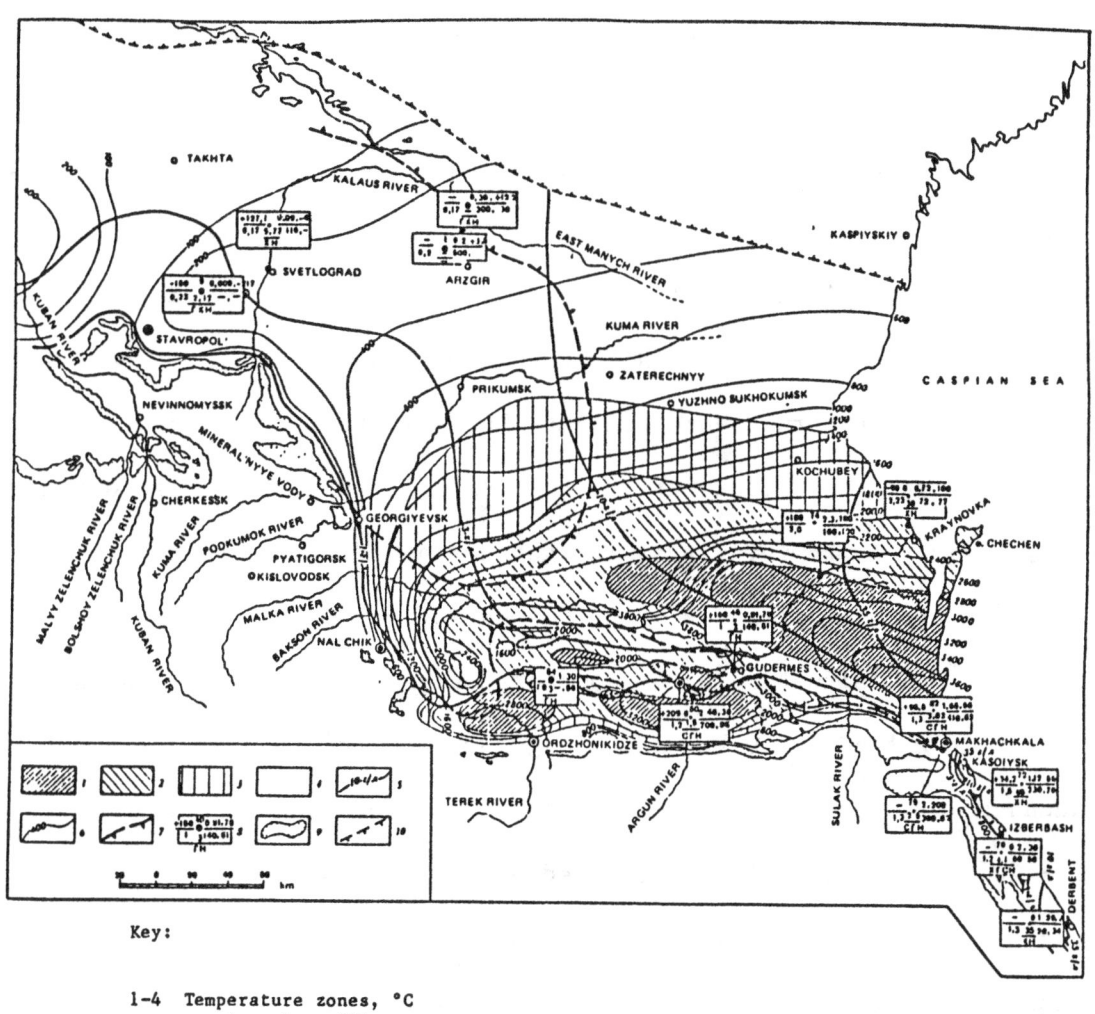

Key:

1-4 Temperature zones, °C
    1 - above 100
    2 - 100 - 75
    3 - 75 - 50
    4 - below 50

5   formation water isominers
6   isoline of bedding depth of Karaganskiy horizon roof, m
7   boundary of water overflowing from Karaganskiy horizon sediments, m
8   hydrogeothermic parameters for areas (wells)
        number above: number of area
        fraction left: numerator, given pressure, meters above sea-level
        denominator: maximum depth to which water-bearing horizon has been
          tapped, thousands of m
        fraction right: numerator (left to right)--maximum flow, thousands of
        m^3/day; excess pressure or depth of static level (minus sign), m; denomi-
        nator (left to right)--maximum production of wells, m^3/day · atm; maximum
        temperature at wellhead, °C; fraction bottom: numerator--overall minera-
        lization of thermal water, g/l; denominator--composition of water with
        respect to dominant ions (C-sulfates, Г-hydrocarbonates, X-chlorides,
        H-sodium)
9   outcrops of Middle Miocene deposits on the surface
10   northern distribution boundary of the Karaganskiy water-bearing formation

Source: [25], p. 141.

high temperature (in excess of 100°C) and pressure. Formed in members of sandstone interbedded with clay, the thickness of the stratal complex reaches 600-700 m. Water-bearing sandstones have high permeability sometimes reaching several darcies.

A number of GWFs in eastern Precaucasia are connected to the Karaganian and Chokrakian water-bearing formations: Makhachkala, Groznyy (Khankal'skaya Dolina), Kizlyar, Izberbash, Ternair, etc. Hypothetical exploitable reserves of thermal water in the Karaganian and Chokrakian formations in eastern Precaucasia are believed to be 6.4 m^3/s.

The Lower Cretaceous stratal complex is distributed over most of Precaucasia. A promising area for geothermal potential mainly in the Lower Cretaceous (Aptian, Albian, Cenomanian and partially Barremian formations) is situated in the southernmost section of Precaucasia and extends in a band 10-50 km wide along the piedmont area of the Greater Caucasus from the Azov to the Caspian seas. Well temperatures of 160-220°C have been noted at the Medvedovskaya, Galugaevskaya and Praskoveyskaya areas. Hydraulic conductivity of the complex in the prospecting zone was about 50 m^2/day; average overflow wellhead production rate was 1,000 m^3/day. Hypothetical thermal water reserves in Lower Cretaceous formations are about 5 m^3/s.

GWFs in Precaucasia (Makhachkala, Groznyy, Izberbash, Cherkessk, Maykop, etc.) are situated in an area contiguous to Meso-Cenozoic outcrops on the northern flank of the Greater Caucasus' meganticlinorium. Characteristics of these GWFs are given below.

The Makhachkala GWF is located near the city of Makhachkala (capital of Dagestan ASSR, population 250,000). The field area is located on a narrow coastal plain along the Caspian Sea. Structurally, the

102

field is confined to the Makhachkala brachyanticline. Commercially viable thermal water sites occur in sandstones of the Karaganian and Chokrakian horizons.* Thickness of the Karaganian and Chokrakian sand and clay deposits at the Makhachkala field reaches 1,350 m. Well no. 160 (interval 1,487-1,518 m) has been tapping the Karaganian horizon for a long time and has produced briney water for over 20 years (water dry residue 3.4 g/l, temperature 62°C) at a rate of more than 1,000 m³/day. Water containing sulfide, sulfate, chloride and sodium water is produced by the B Suite Chokrakian horizon with a dry residue of 5.7 to 10 g/l. Formation temperature is 57-68°C. Thermal water in the B Suite changes significantly throughout the Makhachkala GWF: volume of dry residue ranges from 1.2-3.5 to 11-24 g/l, and formation temperatures range from 59-77°C. Hydrogeothermic parameters of the field are shown in table 9.

In 1964-65, exploratory work was conducted at the Makhachkala GWF in order to make a commercial evaluation of the exploitable reserves of thermal water. In this effort, available abandoned oil wells were re-activated and re-targeted to tap thermal water strata. In December 1966, the State Commission on Mineral Reserves confirmed the exploitable reserves of the Makhachkala GWF as having commercial value and able to produce 6,100 m³/day (about 70 l/sec) if flowing geothermal wells were exploited over a 25-year period.

The Khankal'skaya Dolina GWF (Groznyy GWF) is located in the Oktyabr'skiy administrative region, city of Groznyy (Chechen-Ingush ASSR, population 370,000), and limited to a large anticlinal fold called the Oktaybr'skaya Fold. Slightly mineralized (a few grams per liter)

---

*The Makhachkala oil reservoir was also limited to the Chokrakian horizon; this field began production in the early 1940s and was fully depleted in 1970.

TABLE 9

HYDROGEOLOGICAL PARAMETERS OF GWFs IN EASTERN TRANSCAUCASUS

Location of GWF	Stratigraphic Horizon	Effective Thickness h, m	Permeability, k, darcy	Filtration Resistance Factor $\frac{\mu}{kh}$, centipoise·darcy·cm ·10^{-4}	Filtration Index k', m/day	Piezoconductivity Index a, cm²/sec	Formation Temperature t_f,°C	Overflow Rate of Individual Wells Q, m³/day	Manometric Pressure P_{ms}, atm	Notes
Dagestan ASSR, city of Makhachkala										
4th and 5th settlement area	B Suite, Chokrakian horizon	34-66	1.04-1.71	1.3 / 0.84	1.8-2.9	$1.14 \cdot 10^5$ / $1.84 \cdot 10^5$	55.9-66.0	1950-2320	6.1-7.2	Tests conducted by North Caucasus VNIIgaz and Dagestan Geologic Expedition, 1964-1966
Al'burikent area	C Suite, Chokrakian horizon	35-52	0.4-1.2	2.5 / 0.77	0.98-2.75	$0.58 \cdot 10^5$ / $1.63 \cdot 10^5$	72.0-77.0	605-1080	2.9-4.6	
Ternair area	Member II of Karaganian horizon	28-30	0.42-0.61	3.74 / 2.7	0.81-1.14	10^4-$1.05 \cdot 10^4$	62-64	1175-1470	7.0-9.0	Researched by Dagestan General Geological Expedition, 1966-1967
	B Suite, Chokrakian horizon	34	1.06	1.22	2.05	$0.51 \cdot 10^5$	64	1500	3.15	
	C Suite, Chokrakian horizon	40-59	0.54-0.93	1.40 / 1.19	1.13-1.88	$0.22 \cdot 10^5$ / $0.43 \cdot 10^5$	67-69	864-1230	2.20-4.25	
Chechen-Ingush ASSR										
Khankal'skaya Valley (Oktyabr'skiy area, city of Groznyy)	Strata IV-VII, Karaganian horizon	40	1.3	0.73	3.22-3.70	$0.87 \cdot 10^5$ / $0.93 \cdot 10^5$	89-95	to 2500	3.00-3.35	Researched by Vostokgidrogaz Institute, Saratov, 1966-196:
	Stratum XIII, Karaganian horizon	33-44	0.65-0.80	1.15 / 0.76	2.10-2.37	$0.43 \cdot 10^5$ / $0.63 \cdot 10^5$	98-100	2000	3.5-4.0	

Source: [21].

thermal water circulates in sandstone strata of the Karaganian and Chokrakian horizons. The most productive formations are in the Karaganian formation's IV-VII and XIII* strata, and in Chokrakian's XVI and XXII strata (table 9). Well 3T, which exposed the XXII stratum at a depth of 1,270-1,283 m, produced a water gusher with a flow rate of 3,200 m³/day and a surface temperature of 100°C. Hypothetical exploitable reserves of this field comprise 100 l/sec.

The Kizlyar GWF is located on the Dagestan plain. Kizlyar, population 50,000, is located within the boundary of the field. Two colossal stratal water-pressure systems of eastern Precaucasia--the Apsheronian and Middle Miocene water-bearing pressure formations--were discovered in 1969. Parameters of this field are given in table 10. Commercial reserves at this field comprise 17,000 m³/day (200 l/sec); hypothetical exploitable thermal water reserves in the Middle Miocene deposits of the Kizlyar field are 28,000 m³/day (320 l/s).

The Cherkessk GWF is situated within the boundaries of the northern monocline of the Greater Caucasus, near the city of Cherkessk (center of the Karachayevo-Cherkessk Autonomous Oblast, Stavropol area, population 75,000). Thermal water is confined to sandstaone strata in the Lower Cretaceous deposits (Aptian and Albian stages). Recharge is located 15-18 km to the southwest of the city of Cherkessk, near Ust-Djegutinskaya settlement, at the Lower Cretaceous outcrop belt. Near the city of Cherkessk, Aptian-Albian deposits lie at a depth of 1,100-1,400 m. Thermal water from these deposits near Cherkessk is fresh (mineralization 0.8-1.2 g/l); flow rate is 11-18 l/s; excess wellhead pressure to 20 atm; water temperature at the wellhead is

---

*A large oil field was tied to the XIII stratum, now depleted.

TABLE 10

PRINCIPAL FEATURES OF THE KIZLYAR GWF

Geographic Location of Field	Geologic Age of Producing Beds	Regional Stratigraphic Name of the Producing Beds	Interval Bedding of the Producing Beds from the Earth's Surface, m	Flow Rates of Individual Wells, $\frac{m^3/day}{(million\ gal/day)}$	Water Temperature at Wellhead, °C	Overall Mineralization of Stratal Water, g/l
City of Kizlyar, Dagestan ASSR	Middle Miocene, $N_1^2(ch+kg)$	Karaganian and Chokrakian	2,800-2,900	$\frac{3,000}{(0.8)}$	102	6.6
	Upper Pliocene, $N_2^3 ap$	Apsheronian	1,000-1,200	$\frac{5,000}{(1.32)}$	47	2.0

50-75°C. Hypothetical exploitable reserves of this field are estimated to be 13,000 m3/day.

Caucasian fold system. The thermal water circulation system that includes the Greater and Lesser Caucasus, as well as Talysh, is of the fissure-vein type. Thermal springs are essentially limited to its south-eastern part, and the temperature of these Caucasian springs usually does not exceed 50°C. The highest tempreatures, 64-75°C, were recorded at the Dzhermuk, Isti-Su (Lesser Caucasus) springs, and at a well near Lake Lisi (57°C), in the city of Tbilisi (southern slope of the Greater Caucasus). In the Caucasian fold system, thermal water reserves of the fissure-vein type are estimated to be 1 m3/s, at a temperature of 35-60°C.

In Transcaucasus, stratal thermal water systems are found in the Rionskiy and Kurinskiy artesian basins.

Rionskiy artesian basin. The Rionskiy artesian basin is found in thick Meso-Cenozoic deposits. Commercially exploitable thermal water is found in the Neocomian water-bearing complex. Hydroconductivity ranges from 80-100 to 200-300 m2/day from north to south, where the complex deepens. The renown Tskhaltubo Radon Springs, the Borzhomi Carbonated Springs, and others are derived from Neocomian deposits. A series of wells drilled in Neocomian strata (Zugdidi, Tsaishi, Kindga, Okhurey, etc.) produced flowing thermal water up to 2,000-3,000 m3/day, with temperatures to 100°C. For the greater part of the water-bearing complex, mineralization does not exceed 10 g/1. Hypothetical thermal water reserves at the Rionskiy basin are 2.1 m3/s, with temperatures from 40 to 100°C. Kurinskiy artesian basin. Thermal water in the basin extends across the

Kirovobad area, in a band along the Kura River Valley within the territory of Azerbaijan. The Apsheronian, Akchagylian and Maykop formations, as well as Cretaceous units, offer geothermal potential. Hypothetical exploitable reserves in the Kirovobad area are 2 m^3/s, with temperatures ranging from 40-70°C.

4. Central Asia and Kazakhstan

Central Asia and Kazakhstan occupy an area of more than 1.5 million square kilometers, and are characterized by a diversity of physical, geographic, and geologic features: in addition to the highest mountain ranges (to 6-7 km), deep intermontane and platform depressions have formed here. There are five major artesian basins in Central Asia, as well as the Tajik water-pressure system. In Kazakhstan are the Ust-Urt and Mangyshlak basins (northwestern Kazakhstan), and the Dzharkent artesian basin, which is located near the city of Alma-Ata (south-eastern Kazakhstan).

Central Asia

The principal Central Asian stratal artesian basins are: (1) Syr Darya, (2) Chu-Sarysuy, (3) Amu Darya, (4) Central Kyzylkum, (5) Fergana, and basins in the Tajik water-pressure system. Fissure-vein type thermal water reserves are also found in the Tien Shan and Pamir fold systems.

Potential sites with thermal water are the Syr Darya, Chu-Sarysuy, Amu Darya and Central Kyzylkum basins that produce from Albian-Cenomanian and Turonian-Senonian formations. In the Fergana Basin, thermal water occurs in the Baktriyskiy and Sokhskiy deposits (N_2). Fresh and slightly mineralized water (mineralization to 10 g/l) are found at the basin borders. Temperatures of thermal water in these

basins is 50-70°C (maximum temperature recorded in a well is 86°C).
Geothermal wells often produce well head overflows at 1,000 m3/day and
higher. The Shaulder, Arysskoye, Yuzhno-Kyzylkumskoye, and Tashkent
GWFs produce from the Albian-Cenomanian formations.

Tajikistan. The Tajik water-pressure complex is confined to the Tajik
Depression. Several hydraulically-connected artesian basins produce
from thick terrigenous-carbonate Meso-Cenozoic formations that have
extensive halogen manifestations. Thermal water with low mineralization
levels may be found only on the periphery of these basins; the central
sections contain briney water that is mineralized to 100 g/l. Aptian-
Albian formations in the Surkhan-Darya artesian basin near Dushanbe show
the greatest promise. Here 50-70°C brackish water may be recovered from
wells up to 2,000 m deep.

Tien Shan. In the Tien Shan fold system, fissure-vein thermal water is
widely found, where it surfaces in river valleys. Springs are concentrated on the northern slope of Terskey-Alatau (up to 18 groups), are
found along the depression edge (up to 22 groups), and also in the
central and south-western Tien Shan area (up to 30 groups). Water temperature here reaches 90°C.

Pamirs. The Pamir fold system has also extensive reserves of carbonated
and nitrogenous thermal springs, with mineralization levels usually
below 2 g/l. Hypothetical reserves of thermal water in Central Asia are
13.3 m3/s.

Kazakhstan

Thermal water basins in northwestern Kazakhstan are located in
troughs that are separated by the Mangushlak-Ustyurt dislocation system.
The troughs are formed from terrigenous-carbonate strata of Lower

Jurassic to Tertiary age, and are in excess of 5 km thick. The only promising geothermal water-bearing unit is the Albian-Cenomanian formation. In the Karatau range, fresh and saline water can be found; the briney water has a mineralization level of up to 10 g/l. As the water bearing complex extends to depth, mineralization increases to briney. Flow rates from gusher wells reach 2,500 m^3/day, with water temperature of 55°C.

The Dzharkentskiy artesian basin, situated in south-eastern Kazakhstan near Alma-Ata, contains Meso-Cenozoic terrigenous formations up to 5 km thick. Cretaceous deposits offer geothermal potential; these outcrop in the southern end of the basin, and in the central areas, have been penetrated by wells at depths of 2,500-2,700 m. Water-bearing strata are sandstones with layers of conglomerates. Here fresh and lightly saline thermal water ranges in temperature up to 96°C (at a depth of 2,682 m). Natural flowing rates of wells reach 45-75 l/s. Jurassic and Triassic water-bearing formations also offer geothermal resource potential. Hypothetical overall reserves of thermal water in Kazakhstan are estimated at 6 m^3/s.

5. Western Siberia

One of the largest artesian basins in the world is located in Siberia, a territory of some 2 million square kilometers. Thermal water is confined to Aptian-Albian-Cenomanian and Neocomian formations, which also produce oil and gas to the north of West Siberia.

The Aptian-Albian-Cenomanian water-bearing complex contains terrigenous strata (siltstones, sand, sandstones, and clays) with a thickness of 1,000 m. Stratal water from Aptian-Cenomanian deposits usually has a temperature that does not exceed 60°C. Away from the recharge area,

110

toward the discharge area, the chemical and gas composition of thermal water changes from nitrogenous sodium carbonate water with mineralization of up to 1 g/l in the south of West Siberia (Kolpashevo, Kupino, Ipatovskaya, etc.), to methane sodium chloride saline water with mineralization to 10-15 g/l in the north (Vikulovo, Tara, etc.). Saline water contains high concentrations of bromide (50-80 mg/l) and iodine (10-20 mg/l).

The Neocomian water-bearing formation consists of interbedded siltstones, sandstone, clay, and argillite layers with thickness ranging from 200 to 1,000 m. The depth of the formation changes from 200 m on the edges of the basin to 1,800 m in the center. Thermal water temperature under stratal conditions may reach 95°C. Highest temperatures (75-95°C) were recorded in the Middle Ob River Valley and Irtysh-Tobol'sk interfluvial area. The movement of subsurface water in Neocomian deposits proceeds from the south-east and south in a northern direction. Mineralization changes from 0.3 g/l at the basin periphery to 15-20 g/l for most of the central basin.

Tobol'sk Field. One of the best studied GWFs in Western Siberia is the Tobol'sk field, located within the city of Tobol'sk (population approximately 100,000). Thermal water is confined to members of interbedded sandstones, siltstones, and argillites in the Neocomian strata [48]. The effective thickness of productive sandstones varies from 40 to 110 m, and the formations have good collector properties (permeability exceeds some 1,000 millidarcies). The water is of sodium chloride type, with mineralization of 16-17 g/l, and is saturated with methane (gas factor 0.8-1.0). Excess pressure is 5.8-6.2 atmospheres. Wellhead overflow rates reach several thousand cubic meters per day. Water temperature at the 1,750-1,850 m depth level is in the 74-80°C range. Hypothetical exploitable reserves of thermal water in the Aptian-Cenomanian and Neocomian water-bearing strata in West Siberia are 180 m^3/sec.

6. Southern Part of Eastern Siberia

The southern part of Eastern Siberia is divided into two geologic-tectonic zones: the northern area, the Irkutsk amphitheater which is formed by a wedge-shaped anticlinal nose of the Siberian platform, and the southern Sayano-Baykal fold region.

The Angara-Lena artesian basin is located in the Irkutsk amphitheater. Thermal water with a temperature of 77°C is found here in Ordovician and Cambrian evaporite and terrigenous formations. Thermal water here consists of heavy concentrated brine with a mineralization level of 320 to 500 g/l. Flow rates of wells are usually several liters per second.

In the Sayano-Baykal fold area, Neogene-Tertiary volcanic activity and formation of numerous grabens (the largest of these is the depression taken up by Lake Baykal) relate to a system of large faults that cross the region. Natural outflows of thermal water are focused by fault zones that bound grabens, and are of the fissure-vein type. Within the Sayano-Baykal folded area, 60 types of thermal springs have been located with temperatures ranging from 20 to 80°C. Springs are especially numerous in the Baykal rift zone, and are generally fresh water, nitrogenous, sodium sulfate-hydrocarbonate thermae. Flow rates fluctuate from several liters to a few dozen liters per second. When exploratory work is carried out at sites adjacent to the springs, flow rate and water temperature may be substantially increased

112

(Goryachinskoye, Pitatelevskoye, Nilovo-Pustynskoye and other fields).
According to data from the AN SSSR (Siberian Division) Earth Crust
Institute, the overall flow rate of all known thermal springs in the
Baykal fault zone is 1 m^3/s. In addition to fissure-vein type thermal
water sources, stratal water systems are widely found in the
Selenginskaya and Tunkinskaya intermontane depressions of the Baykal
folded system. In these depressions, slightly mineralized water (to
3 g/l) with temperatures to 75°C is found in Miocene-Pliocene deposits.
The wells are of the naturally flowing type. Hypothetical exploitable
thermal water reserves in these two depressions (5,000 sq. km in area)
are 1 m^3/s, with a temperature ranging from 35-40 to 70-75°C.

7. Transbaykal and Amur Areas

The Transbaykal and Amur fold zone is the eastern branch of the
Asian Hercynian Belt. In this zone, few of the thermal springs confined
to fault zones and peripheral to intermontane basins have practical
value. Mineralization of these springs is below 1 g/l, and production
rate does not exceed several liters per second at a temperature of
35-72°C. However, as a result of exploratory efforts, more substantive
thermal water resources have been located: at the Kuldurskiy Spring,
which had a natural flow rate of 1.6 l/s, a flow rate of 22 l/s and
a temperature of 72°C was produced from two wells. Hypothetical
exploitable thermal water resources in the Transbaykal and Amur area are
estimated to be approximately 0.1 m^3/s, and a temperature of 40-70°C.

8. Northeast and Chukotka

This territory, covering an area of 4 million km^2, is a Mesozoic
fold belt. On the west, this zone is bounded by the Priverkhoyanskiy
foredeep; the Chukotsko-Kataziatic volcanic belt forms the southern

boundary. There are 5 principal geologic and structural features in
this area: 1) Kolyma fold belt; 2) Yano-Kolyma fold belt; 3) Chukotka
fold belt; 4) Okhotsk fold belt, and 5) the Okhotsk-Chukotka volcanic
belt (figure 13). The relief of the territory is complex and varied--
from mountain ranges more than 2,500 m high to lowlands only slightly
above sea level. This region of severe climatic conditions is almost
entirely located in permafrost that is broken up only under large river
and lake valleys. Thickness of the permafrost layer increases south to
north, from 100 to 500 m.

FIGURE 13

GEOSTRUCTURAL SUBDIVISIONS OF THE NORTHEASTERN USSR

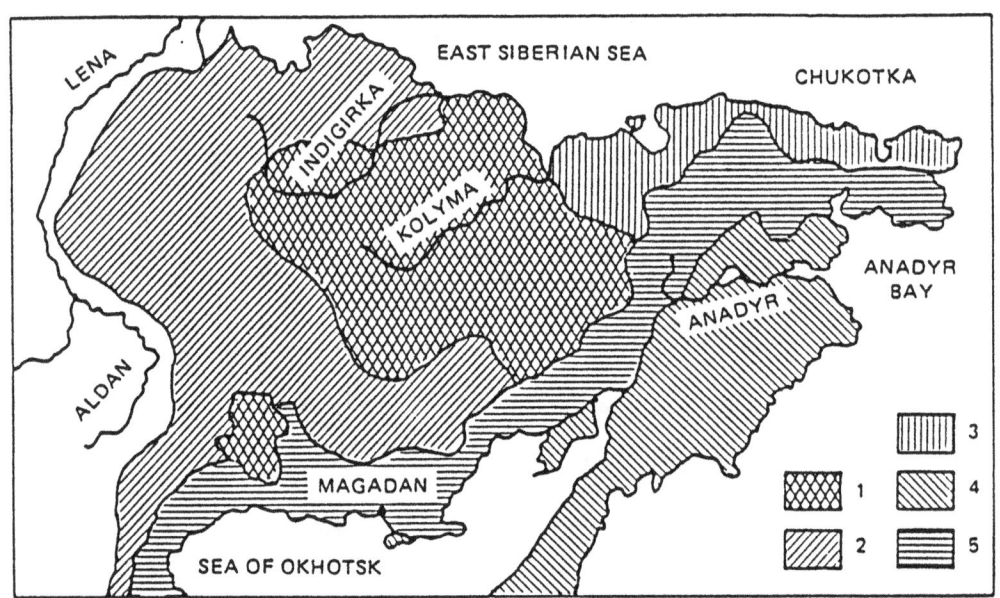

1   Kolyma fold zone
2   Yano-Kolyma fold zone
3   Chukotka fold zone
4   Okhotsk fold zone
5   Okhotsk-Chukotka volcanic belt

Source: Teplo Zemli i ego izvlecheniye. Kiev, Naukova Dumka, 1974, p. 229.

With respect to geothermy the vast territory of the Northeast
is poorly known, and underground temperature distributions are still
largely hypothetical. The extent of the thermal flow in the Northeast,
based on AN USSR data, is assumed to be 1.42 $\mu cal/cm^2 \cdot s$. Anomalous
values of the geothermic gradient (to 40°C/km) have been noted in the
superimposed Cenozoic basin of the Kolyma massif. It is believed that
pressurized thermal water with temperatures to 80°C are found in large
artesian basins (Yano-Indigirskiy, Kolyma-Indigirskiy, etc.), at depths
to 3,000 m.

The northeastern territories of USSR are believed to offer practical, commercial quantities of thermal water, however sufficient data
are as yet unavailable [50].

The area whose geothermal potential is known is the Okhotsko-
Chukotskiy volcanic belt. The high geothermal pressures and abundance
of dislocations with continuity breaks have resulted in fissure-vein
thermal water reserves. In this region, 19 groups of springs with temperatures to 90°C have been studied. The Chukotka Peninsula has the
highest number of thermal sources. Flow rates of individual clusters
reach 10-20 l/s and more at temperatures of 60-80°C (Chaplinskiye,
Lorinskiye, Sinyavinskiye, and other springs). Mineralization is
generally low, 3-4 g/l.

Overall hypothetical reserves of thermal water for the Chukotka
Peninsula are 0.2 m^3/s, with a temperature of 40-80°C. In the
Okhotsk area, there are a number of thermal springs with temperatures
from 35 to 90°C and flow rates of 15 l/sec. The Tal'skiy GWF in this
area has a temperature of 90°C, flow rate of 10 l/sec and mineralization of 0.4 g/l. In the Okhotsk region, overall hypothetical reserves

of thermal water are 0.1 m^3/sec with water temperature from 35-40 to 90°C.

In the Maritime sector, only 5 freshwater nitrogenous thermal springs have been recorded. Forecasted thermal water reserves in this area are 0.05 m^3/s.

9. Kamchatka

The Kamchatka Peninsula is part of the Asian-Alaska sector of the Circum-Pacific Cenozoic volcanic belt. The modern tectonic activity in Kamchatka reflects the dominant role played by northeastern dislocations, which determine volcanic activity and hydrothermal phenomena. All GWFs and hydrothermae are of the fissure-vein type. Away from the tectonic dislocation areas, rocks forming the fields are practically waterless. Overall mineralization for thermal water here is usually insignificant, in the range of 1-2 g/l and 2-4 g/l for the vapor-hydrothermae. Numerous groups of 40-100°C thermal springs are located in the Eastern volcanic zone and in the Central Kamchatka uplift. GWF exploration has shown that the reserves in these fields substantially exceed natural discharge.

Those fields with low or average potential are the Paratunka, Avachinskaya Bay (south shore), Malkinskoye, and Nalychevskoye GWFs. Hypothetical reserves of thermal water with temperature to 100°C are forecast at 2 m^3/s.

A number of superheated water fields have been discovered on Kamchatka, with an average calorific value of 150 Kcal/kg. Main fields of this type have been combined into 3 groups: Pauzhetka (Pauzhetka, Koshelevskoye GWFs), Uzono-Semyachinskoye (Nizhne-Semyachinskoye, Verkhne-Semyachinskoye, Uzonskoye GWFs); and Mutnovsko-Zhirovskoye

(Nizhne-Zhirovskoye, Verkhne-Zhirovskoye, Severo-Mutnovskoye GWFs). In addition, at the Bolshe-Bannoye field, 165 kg/s of steam-water mixture with a calorific value of 150 Kcal/s was produced.

Overall reserves of steam-water mixture in Kamchatka have been estimated at 4-5 t/s, which corresponds to 350-400 MW of power [16]. According to Vakin, energy resources of the main Kamchatka vapor-hydrothermae fields represent 411 MW [54].

The Pauzhetka GWF is located in the southern part of Kamchatka, 230 km south of the city of Petropavlovsk-Kamchatskiy, in the Pauzhetka River basin. The field is a part of a large hydrothermal system, one that joins the thermal fields of Kambal'nyy Range and the Pauzhetka thermal field. It is confined to psephite tuff, lying at a depth of 40-90 to 270-350 m. Maximum temperatures of 194-200°C in water-bearing rock interval and the underlying tuff-breccia were recorded in the south-east sector. In the tuff mass, superheated water circulates in pores and along fissures in a droplet-liquid state. Most of the thermal springs at this field are characterized by low flow rates. The flow rate of the largest boiling spring (Paryashchiy-1) reached 10 l/s. At the Pauzhetka field, there were also two small geysers--Liliput and Lenivyy, which shot jets of water to a height of less than 1 meter [51]. Water in the springs and the geysers is siliceous sodium chloride type with an overall mineralization of 3 g/l; with respect to gas, it is of carbon dioxide type.

Temperature anomaly in the discharge area of the Pauzhetka thermal springs is induced by motion of hydrothermae. Exploration of the field was conducted by drilling 21 boreholes (1958-1965). During individual flowing tests at a pressure of 2 atm, well yield fluctuated widely, from 2 to 33 kg/s. Exploitable reserves of the field, which were estimated on the basis of experimental-exploitable flowing test data collected over a year, were 124 kg/s of steam-water mixture, and average enthalpy of 170 Kcal/kg. This was equal to the thermal power of 2,100 Kcal/sec (the GKZ accepted this figure to indicate exploitable [category B] reserves); 44.9 kg/s were confirmed as category C_1 reserves. Known reserves assure the operation of the 5,000 kW electric power plant.

The Paratunka GWF is situated in one of the most populated regions of Kamchatka, 25 km southwest from Petropavlovsk-Kamchatskiy in the Yelizovo Settlement region. The field is located in a small intermontane artesian basin that fills the complex graben structure of the Paratunka River valley. On the surface, the field is marked by groups of thermal springs that appear in flood plains and along the slopes of the Paratunka river valley and its tributaries. However, the extent of the observable natural discharge of thermal water is insignificant because the principal mass of water remains in alluvial strata, becomes impoverished, cools, and is carried away by groundwater underflows.

The regime of the field's thermal water is connected to surface water in the area. Fissured tuff, andesites, basalts, and Paratunka and Alneyskaya series tuff sandstones that lie in the graben serve as water-bearing strata. The field is broken up by a system of faults to resemble a "broken plate" structure.

At the Lower Paratunka section, water inflow zones are distributed unevenly along the wellstem: the deepest water inflow zone has been located at 1,338-1,345 m. With respect to gas, nitrogen predominates (96% of volume). Mineralization of the thermal water is within the 0.9-2.1 g/l range. Cation and anion composition of the water varies from calcium-sodium sulfate to sodic-calcium chloride-sulfate types.

118

Experimental-exploitable flowing tests were performed at 40 wells selected on the basis of productivity.

Exploitable reserves of thermal water at the Lower, Middle, and Northern Paratunka sections were confirmed by GKZ in April 1969 to be 270 l/s, with an average temperature of 77°C.

The Paratunka GWF is one of the largest explored medium-potential thermal water fields in USSR. Based on exploratory and development drilling and test efforts, it was established that Paratunka GWF reserves may be increased by an additional 250 l/s, with a temperature of up to 75°C (forecast reserves of the Upper Paratunka and Karymshinskoye areas).

10. Kurile Islands

Like Kamchatka, the Kurile Islands are a region of modern volcanic activity. Of practical interest for thermal water production are five of the largest islands--Paramushir, Simushir, Urup, Iturup and Kunashir. A number of active volcanoes are found on these islands, as well as numerous thermal springs and steam jets. The Kuriles, like Kamchatka, are formed from Neogenic volcanic-sedimentary deposits that are broken up by a number of major dislocations that focus thermal activity. On these islands, hypothetical exploitable thermal water reserves at temperatures to 100°C equal 1 m^3/s; vapor-hydrothermae reserves of the Kurile Islands are estimated to be 1 t/s. To date, however, only the Goryachiy Plyazh thermal springs field has been explored on the Kurile Islands.

The Goryachiy Plyazh GWF is located on Kunashir Island on the shore of the Pacific, near the Mendeleyev Volcano, approximately 8 km from the city of Yuzhno-Kurilsk. Surface manifestations of thermal

water include boiling springs and dispersed-steam areas. These have
been observed in a kilometer-long strip of ocean shore, up to 200-250 m
wide. Based on thermometric data, areas with temperature from 30 to
100°C at a depth of 15 cm make up 5,000 sq. meters. Spring flows are
low and generally do not exceed 0.3 l/s. The field is formed in
Neogenic tuffs, tuff-sandstone, tuff-conglomerate, and Tertiary (re-
deposited volcanic rock) formations. The Neogene layer is 570 m thick;
bedding within the field is nearly horizontal.

Superheated water is derived from Upper Miocene fissured volcanic
formations, and is lightly mineralized, sodium chloride water. Carbon
dioxide and hydrogen sulfide dominate the gas fraction. Well flows are
11-30 kg/s, with enthalpy of 173-265 Kcal/kg; wellhead pressure 1-1.5
atm; stratal water temperature at a depth of 130-150 m is 150-170°C.
Reserves of water-steam mixture at the field have been confirmed by the
GKZ at 3,500 t/day (40 kg/s).

11. Sakhalin Island

A series of artesian basins are located on Sakhalin Island:
Severo-Sakhalinskiy, Paronayskiy, Susunayskiy, and Tatarskiy.

The Severo-Sakhalinskiy (North Sakhalin) artesian basin is best
known because most exploratory and prospecting oil wells are concentrated here. The basin area is 22,000 km^2, and the majority of
water-bearing formations are Neogenic terrigenous sand, sandstones, and
siltstones.

Thermal water temperature of this complex is usually in the 60-80°C
range, and stratal water mineralization is usually low, to 3 g/l
increasing with depth to 20 g/l. The highest temperature, 97°C, was
recorded at a depth of 3,300 m (Tungor). Wells overflow naturally.

120

The Susunayskiy Intermontane basin, of 1,500 km2 contains sedimentary Neogenic formations with thickness of up to 3,000 m. Based on data for exploratory oil wells, the Lower Miocene thermal water-bearing formation is of special interest. Mineralization of stratal water is 8-9 g/l, and temperature is 70°C at a depth of 1,800 m (Dolinskaya area).

The Tatarskiy Artesian Basin occupies an area in excess of 8,000 km2. Neogenic and Paleogenic deposits that form the basin reach a thickness of 8 km. The deposits are enriched by tufogenic material. Stratal-fisure and fisure-vein water systems are widely found. Paleogene formations near the faults have good permeabilities. An increased geothermic gradient (up to 4°C/100 m) is typical for the Tatarskiy artesian basin. An absolute temperature maximum of 128°C was found in Upper Miocene deposits in the Krasnogorskaya area at the 2,808-2,945 m interval. Hypothetical thermal water reserves comprise 3 m3/s.

Overall hypothetical thermal water reserves in stratal type basins in the USSR are 232 m3/s, with temperatures of 40-100°C. In fissure-vein systems, hypothetical thermal water and steam-water mixture reserves are estimated to be 6.6 m3/s and 5 t/s, respectively [16].

CHAPTER IV

PROBLEMS AND PROSPECTS FOR GEOTHERMAL ENERGY USE

Despite the many obvious advantages of geothermal energy--its
renewability and limited environmental impact to name but two--many factors continue to impede its use. Prior to the 1960s, when expertise
on development and use was still in its infancy, hydrothermal resources
had not yet been characterized extensively; moreover, the necessary
technical and economic parameters governing rational use were lacking.
The substantial geological exploration and research efforts conducted in
the Soviet Union in the last 20 years eliminated many of these obstacles. At the same time, some problems which inhibit commercial use
remain unresolved: these include limited transportability (20-30 km at
the outside), technological difficulties (intensive corrosion and
mineral deposition from saline and briney water in pipelines and well-
stems) and disposal of "waste" water, and the need to reconcile these
factors with regional climatic, geographic, and economic features. In
this chapter, the current state of and future prospects for Soviet
geothermal energy use will be discussed by resource type.

1. Thermal and Superheated Water

As a result of the efforts undertaken to develop the geothermal
exploration program, the practicality, effectiveness, and commercial
competitiveness of thermal water in comparison with other energy sources
has been proved in a number of instances, and the technical problems of
exploring, developing, and exploiting GWFs have been essentially

resolved. In 1979 in the USSR, thermal and superheated water energy provided heat for 4,500 apartments, and hot water for 300,000 people, heated 50 hectares of hothouses, and generated 16 million kWh of electricity. In 1980, the Soviet Union was exploiting 36 GWFs comprising 170 wells for a total of 40 million m^3 of hot water [60]. Steam production, which occurred only at the Pauzhetka power station, was 250-300 thousand tons annually during the period 1968-1980.

Although they fell somewhat shy of plan targets [8], these figures represent but a fraction of hypothetical Soviet thermal and superheated water reserves. Estimates of hypothetical thermal water reserves show the overall energy of these reserves to be on the order of 200 million Gcal annually [16]. Not surprisingly, the USSR has continually increased its planned thermal water recovery targets (figure 14).

This greater effort will be augmented by an expanded R&D program directed toward enhancing understanding of hydrothermal basins and GWFs and developing and exploiting new, promising sites, especially those in Kamchatka, the Caucasus, Western Siberia, and Central Asia. There is also significant potential in Kazakhstan (Mangyshlak Peninsula), Sakhalin, the Transbaykal area, and the Kurile Islands. Plans foresee the study of thermal and superheated water for use in power production and greenhouse and hothouse farming in Kamchatka and the Kurile Islands, for providing heat to cities, and accelerated development of operations in Siberia, the Far East, Northern Caucasus, and Transcaucasia [61]. Also planned are prospecting efforts in the area of the Baykal-Amur Railroad, beginning first in the Charskaya, Muyskaya, and Upper Angara Depressions [62]. In the expanded geothermal program, priority has been accorded to Kamchatka and the Kuriles; the main body of research is to be concentrated on the Mutnovsko-Zhirovskaya and Nizhne-Koshelevskaya areas

123

FIGURE 14

INCREASE IN THERMAL WATER PRODUCTION IN THE USSR

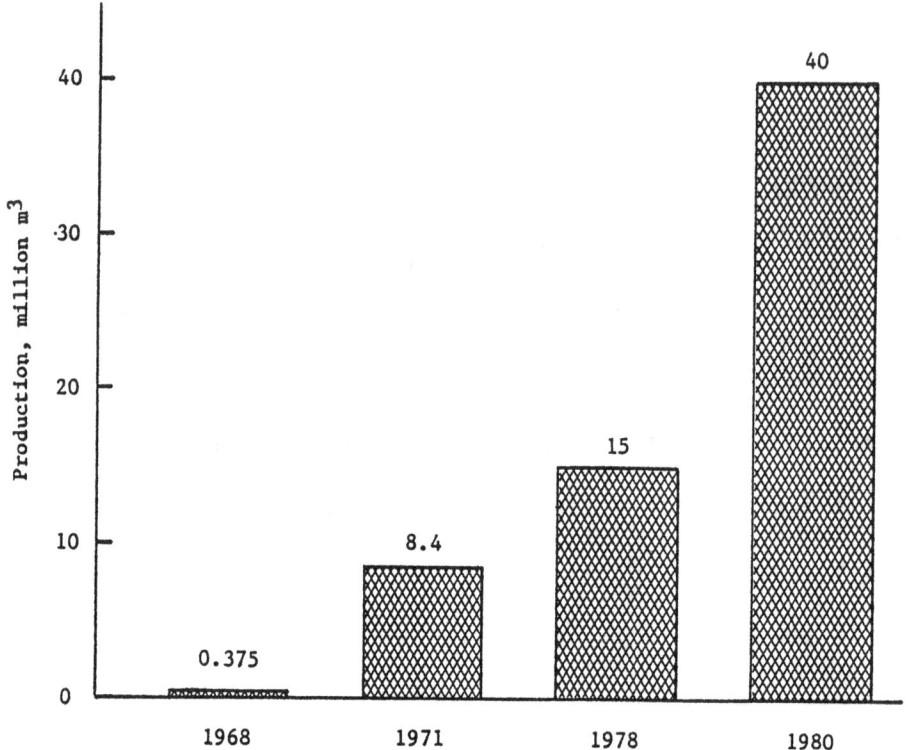

Source: Nurshanov, V. A., and B. M. Drozd. Ispol'zovaniye glubinnogo tepla
Zemli v narodnom khazyaystve. In: Izucheniye i ispol'zovaniye
glubinnogo tepla Zemli. Moscow: Nauka, 1973. Gadzhiev, A.
Skovannaya energia. Pravda, July 5, 1981.

of Kamchatka. Before proceeding to a discussion of specific types of

thermal and superheated water exploitation, however, attention ought

first be turned to those technical and technological issues which pre-

sent problems that apply across the board for the use of these resour-

ces.

Recovery and Disposal of Thermal Water

At present, Soviet thermal water recovery is conducted via natural

formation pressure. In many cases, thermal water cannot be used imme-

diately in heat supply systems as it quickly coats pipelines and forms

deposits in well stems (e.g. in Bol'she-Bannoye, Kizlyar, Nalchik, etc.). For example, at well 5-T at the Kizlyar GWF, which had a production rate of 3,000 m3/day (wellhead water temperature 100°C, overall mineralization 8.6 g/l) the 100 mm discharge pipe was almost fully clogged with calcium carbonate deposits after only one and a half months. The most extensive deposits were observed in sections with the highest pressure fluctuations (at pipeline elbows, flow regulators, and valve areas). This mineral deposition results from disturbances in the water's carbonate balance caused by the release of dissolved carbon dioxide. Stabilization of thermal water at well 5-T, which was used to supply heat to the new Cheremushki district (mikrorayon), was facilitated by a vortex reactor [63]. When minerals are deposited in the well stem, plugs must be either drilled through or dissolved with the aid of sulfuric, chloride, and boric acids. Use of these agents, however, substantially reduces the exploitation efficiency of the GWF.

Another serious problem at commercial geothermal sites is metal corrosion. Corrosive qualities of geothermal water are determined by the presence of such components as O_2, S^{2-}, Fe_2O_3, CO_2, Cl^-, and SO_4^{2-}; depending on the concentration of such substances, the corrosion process occurs differently for different metals. Corrosion may be reduced by adding inhibitors and agents that increase pH levels, as well as the use of heat exchangers, special materials and coatings, and preliminary deaeration of thermal water. At the Pauzhetka GeoTES, equipment corrosion was reduced to insignificant levels through the use of carbon steel, which is corrosion-resistant, as well as by treating the water with inhibitors and stimulators. At the Nalchik Resort in the Northern Caucasus a double-pipe heat exchanger is used; the mineralized water transmits a portion of its heat to fresh water, thus limiting the extent of corrosion and mineral deposition. Technical and economic studies have shown, however, that deaeration of thermal water is the most effective approach to alleviating these problems.

Thermal water may also contain high concentrations of toxic substances. For example, at the Paratunka GWF, arsenic concentration (0.5-0.6 mg/l) prohibits the use of this water for household hot water purposes. Research is currently underway to develop a system which removes arsenic from the thermal water by using ion-exchange tars and inorganic ionites.

The problem of environmentally sound disposal of used water also inhibits the exploitation of geothermal water; disposal presents particular difficulties when the water is highly mineralized (exceeding 10 g/l). Expelling used thermal water into natural waterways, evaporation ponds, and storage facilities usually requires large capital expenditures and does not address the problem of protecting surface and subsurface water supplies. Alternatives include pumping used water into deep, securely sealed strata, and returning it to the stratum from which it was extracted. Although the latter method usually requires a significant amount of energy--particularly when deep-lying strata are involved--the returned water aids in maintaining formation pressure and in some cases can even make exploitation commercially viable. Despite the fact that this method is only advantageous when the given strata are highly porous and water-conductive, the USSR favors it whenever possible in oilfield and commercial geothermal site development.

Conversion to Electric Power

Electric power generation is one of the most effective uses of geothermal heat. Current technology makes it possible to obtain more than 20 kWh of electric energy from 1 ton of water with an initial temperature of 150°C [57]. Another important feature of geothermal electric power plants is their constant power production throughout the entire year, and in this regard they do not differ from their nuclear and thermal counterparts. Moreover, exploitation of GeoTES (geothermal electric power plants) requires no outlay for fuel since reserves of geothermal water are renewable, and the capital costs for construction are lower because the need to build service roads for fuel deliveries, boilers and smokestacks, etc. is eliminated.

The outlay for the construction and exploitation of a GeoTES includes the costs of geologic-exploratory operations, borehole drilling, turbines, pipelines to carry steam, separators to remove steam from steam-water mixture, facilities for extracting associated components from turbines, etc., and the operating expenses for the plant and commercial facilities. Capital investments and amortization deductions represent 70 percent of the overall costs. Optimal technical and economical parameters of a GeoTES are established by analyzing the thermodynamic, construction, and geographic factors.

The technical and economic analysis of the only currently operable Soviet GeoTES--at the Pauzhetka superheated water field--shows that the 5MW geothermal power plant was superior to a fossil-fuel electric plant of the same capacity. The net cost of power generated by the plant's two 2.5 MW condenser turbines is 0.75 kopecks/kWh, with an annual production of 25 million kWh. GeoTES projects are now underway in the Stavropol area, Dagestan, and Transcaucasia. For example, in 1983 plans required completion of the study of the Mutnovskoye formation and hydrothermae in order to assure priority construction of the first phase of a 50,000 kW power plant and to continue the study of the formation in order to build two subsequent phases of 200,000 and 400,000 kW, respectively.

Such projects notwithstanding, electric power from heated water resources is destined to remain on the periphery of the Soviet geothermal program. Since the majority of the USSR's thermal water reserves (over 90 percent) consist of low- and medium-potential thermal water (temperature to 100°C), they are unsuitable for conversion into electric power. Efforts to convert the heat in thermal water cooler than 100°C into electric power using a secondary heat carrier have failed to render such power generation commercially feasible. Moreover, those formations which yield steam with characteristics suitable for electric power generation using available technology are concentrated in Kamchatka and the Kuriles, far from population centers.

Direct Uses

As mentioned above, 90% of hypothetical Soviet geothermal water reserves have temperatures of 40-100°C. Taken together with the fact that these reserves are found over an area that exceeds 20 percent of the country, the most practical use of Soviet geothermal water is thus in geothermal space-heating systems. The direct use of thermal water for hot water and for industrial processes also makes it possible to save fuel resources as well as to conserve other water supplies. Highly mineralized thermal water (to 10 g/l) may be used for limited hot-water usage, such as for household needs. The main specific feature of ther-

128

mal water in low temperature heat supply is that it is used once at a constant temperature (as opposed to a usual heat carrier, which is returned for reheating after initial use, and upon which climatic factors have much greater influence). In one of the first papers devoted to the issue of effective utilization of thermal water for space heating [19], a comparison of technical and economic indicators of heat supply in a mikrorayon (residential complex) in the city of Groznyy was carried out. For the 4,000 residents, heat was supplied by a local boiler system that ran on imported coal and by GWF (temperature 90-95°C, overall mineralization 1-2 g/l, hardness 3-4 mg equivalent). Here the system using geothermal water was shown to be more advantageous. The most detailed thermal and economic analyses of geothermal heat supply for the different economic regions in the USSR were carried out in the 1960s by TsNIIEP's Engineering Equipment section under an agreement with USSR Mingazprom. Analysis of such systems showed that given identical parameters for a geothermal energy source, results vary widely depending on the system selected [64].

Aside from electric power production, the highest efficiency in geothermal water use is in hot water supply systems; this efficiency may be achieved via special low temperature systems and the use of additional heating and thermal pumps, as well as recirculated water for other consumers. Studies for a series of GWFs (Paratunka, Makhachkala, Ternair, Izberbash, Cherkessk, Zugdidi, Tsapshi, Mendzhi, Astara, Massaly, Pitatelevskoye), which involved a broad range of hydrogeothermic and climatic conditions, technologies, and users, have shown that the efficiency coefficient for geothermal hot water supply systems fluctuates within 0.3-0.8, while for forecasting purposes, it may be assumed to be equal to 0.55.

The estimated thermal capacity of the geothermally-heated water supply and fuel savings for varying regions of the USSR was defined in 1980; it is presented in table 11. Calculations have shown that under typical conditions for heat supply (except for Kamchatka and the Kurile Islands), fuel savings could have reached approximately 110 million rubles in 1980 [16].

The pricing structure in effect in the 1960s and early 1970s showed many exploratory-prospecting wells (those deeper than 2,000 m with daily production below 1,000 m^3) to be unprofitable. At the same time, however, in many cases exploitation of geothermal fields in the USSR has confirmed the effectiveness of this energy resource for heat and hot water supply. In the 1970s, in regions with promising potential for using thermal water, 50% of the thermal load for the community and household sector (150 million Gcal/year) was covered by using small boilers and individual networks. For the rayon (county) of the city of Groznyy, which was provided with inexpensive tap water and fuel (natural gas), heat provided by geothermal wells costs 1.5 times less than other readily available energy sources supplied by the state. In the Makhachkala, geothermal heat costs consumers less than its counterpart from Dagenergo (Dagestan Power Co.), by a factor of greater than three. On Kamchatka, the cost of geothermal energy is below thermal energy costs in the Petropavlovsk-Kamchatskiy heat supply network by 5.5 times for the general population, by 8 times for agriculture and by 17 times for commercial users. Thus the use of geothermal water for heating and hot water supply is economically very justifiable.

Due to the wide distribution of geothermal water having a temperature of 40 to 60°C (reserves of this water are estimated to be

TABLE 11

RECOMMENDED ANNUAL HEAT PRODUCTION, ESTIMATED LOAD OF CONSUMERS
AND FUEL ECONOMY BASED ON WATER TEMPERATURE GRADATIONS
IN PROMISING REGIONS

Region	Temperature of Thermal Water, °C	Annual Production of Heat to Consumers ($Q_{year}^{factual}$), Gcal/yr (in thousands)		Potential Estimated Load Users, Gcal/hr (in thousands)		Potential Annual Savings in Fuel, Conventional Fuel/yr (in thousands)	
		Overall Predicted Resources	For 1980	Overall Predicted Resources	For 1980	Overall Predicted Resources	For 1980
West Siberia	40-60	122,000	25,000	31.4	6.5	23,200	4,800
	60-80	31,000	6,000	8.0	1.5	5,900	1,150
Transcaucasus	40-60	17,000	8,500	4.4	2.2	3,250	1,620
	60-80	7,700	4,400	2.0	1.1	1,470	840
	80-100	4,300	2,150	1.1	0.55	820	410
Kamchatka	40-60	420	340	0.11	0.087	80	65
	60-80	1,100	880	0.28	0.23	210	170
	80-100	720	570	0.18	0.15	140	110
	100-200	9,800	7,300	2.5	1.9	1,870	1,400
Kurile Islands	60-80	550	440	0.14	0.11	110	84
	80-100	720	570	0.18	0.15	140	110
	100-200	2,450	1,950	0.63	0.5	470	370
Sakhalin	40-60	640	250	0.16	0.064	120	48
	60-80	550	330	0.14	0.085	105	63
Eastern Siberia and some regions of the Far East	40-60	2,500	1,000	0.64	0.26	480	190
	60-80	2,200	880	0.56	0.23	420	170
Azerbaijan SSR	40-60	2,100	500	0.55	0.13	400	95
	60-80	550	330	0.14	0.085	105	63
Georgian SSR	40-60	1,300	1,000	0.33	0.26	250	190
	60-80	1,100	880	0.28	0.23	210	170
	80-100	1,450	570	0.36	0.15	280	110
Kazakh SSR	40-60	11,900	2,500	3.0	0.64	2,270	480
	60-80	1,600	990	0.41	0.25	310	190
	80-100	2,400	860	0.61	0.22	460	160
Kirghiz SSR	40-60	1,700	500	0.44	0.13	320	95
Tajik SSR	40-60	1,300	500	0.33	0.13	250	95
	60-80	550	220	0.14	0.056	110	42
Uzbek SSR	40-60	6,250	1,300	1.62	0.33	1,200	250
	60-80	550	550	0.14	0.14	105	105
Ukrainian SSR	40-60	1,700	700	0.44	0.18	320	130
Total for temperature gradations	40-60	167,810	42,090	43.42	10.911	32,140	8,058
	60-80	47,450	15,900	12.23	4.046	9,055	3,047
	80-100	9,590	4,720	2.43	1.22	1,840	900
	100-200	12,250	9,250	3.13	2.4	2,340	1,770
TOTAL		237,100	71,960	61.21	18.577	45,375	13,775

Source: [64]

201 m3/s, i.e., 80 percent of the overall quantity of hypothetical Soviet geothermal water reserves), this water is viewed as a rather promising source of heat for covered production of vegetables. Greenhouse and hothouse raising of vegetables may efficiently use water with temperature fluctuations of as much as 35°C. The method of supplying heat to greenhouses is selected in accordance with a number of features: the geothermal source, solar radiation, temperature difference, etc. These factors influence not only the type of system used, but also the heat carrier used in it. Forced air heating has been shown to be the most economically sound method. Another important variant involves the use of supplementary boiler installations which employ low-potential thermal water for covering short-term loads.

The following minimum requirements are stipulated for geothermal sources to be used for heating greenhouses and hothouses:

a) the thermal capacity of geothermal heat source must be at least 2.5-3 Gcal/hour;

b) the temperature of water rising to the surface may be no less than 40°C;

c) excess pressure at the wellhead must be at least 1-1.5 atm;

d) overall mineralization of the water may not exceed 25 g/l [65].

Local geologic and economic conditions must of course also be taken into account in evaluations of the potential of geothermal resources for agricultural uses. All told, heating hothouses and greenhouses has proved well worthwhile. For example, near the Ternair GWF (Dagestan), winter greenhouses heated by 60°C geothermal water yield some 19,000 rubles additional annual income. The Soviet Union intends to explore geothermal sources sufficient to assure the expansion greenhouse and hothouse farming from 36 to 120 hectares.

Balneologic use of thermal water in the Soviet Union will continue to grow. The most important factors that determine the potential of using thermal water for medical purposes are the temperature of the natural water and the quantity and composition of the dissolved salts and gases. Water with a temperature range of 35-42°C is applicable for balneologic purposes, and 20-35°C water can be used in swimming pools.

Nitrogenous, lightly mineralized alkaline thermal water which has diverse ion composition and a high silicic acid content (in excess of 50 mg) has medicinal properties, and is used at many Soviet resorts for balneologic purposes. (It is also employed for heating medicinal water facilities, hothouses and greenhouses, and for household purposes at Tsaishi, Obigarm, Tashkent, Talaya, Nachiki, Kuldur, Goryachinsk, etc.) In the early 1970s, the overall quantity of thermal water used by resorts and balneologic facilities in the USSR comprised some 1.7 m^3/s.

2. Hot Dry Rock

Exploitation of the conductive heat resources in the rock which forms the Earth's crust would vastly increase not only the volume of geothermal energy recovered, but would also greatly expand the geographic range in which geothermal energy could be exploited. Moreover, the thermal energy of hot dry rock (HDR) is extremely high: for example, the energy released by cooling 4 km^3 of rock from 350 to 177°C would equal that produced by some 48 million tons of oil [7]. Despite the fact that the USSR possesses considerable geothermal reserves in the form of HDR, several factors continue to impede their exploitation.

One such factor is accessibility. In areas with high positive geothermal anomalies, HDR with a temperature of 250-350°C is located at accessible depths; for example, on Kamchatka and the Kurile Islands, superheated rocks lie at depths of some 2-3 km. Such areas occur relatively rarely, however. In regions with normal and reduced geothermic gradients, superheated rock is found at depths of 10-15 km, where permeability is very low due to the extreme compactness and recrystallization of the grains from which the rock is formed.

Exploitation of HDR reserves is also limited by current technology. Despite ongoing research efforts, highly efficient means of directly extracting thermal energy from HDR have yet to be developed. Turbines have been used to transform the energy of expanding steam into electrical power, but they have low efficiency.

In an attempt to circumvent these problems, some current projects are directed toward the creation of closed circulation systems in which a medium--such as water-steam mixture--is artificially introduced into the strata, heated, extracted, and returned to the stratum. An advantage inherent in this approach is that artificial circulation systems could be created practically anywhere using previously set heat carrier parameters. Those projects would employ either natural or artificial "underground boilers" [50]. The former consist of isolated geostructures (or segments of such structures) or zones with anomalously high formation pressures within whose boundaries a closed circulation system could be created for the heat carrier, without substantial heat carrier losses. The latter implies the creation of such systems in large, previously disturbed HDR areas. Explosions could be used to activate unproductive natural thermal basins as well as to create large fissure zones in extensive HDR formations [66]. Explosions of the needed magnitude would likely be nuclear; as a result, R&D on HDR exploitation is sensitive, and information is thus not readily available.

The development of high temperature underground boilers will be realized only if a number of technical, technological, seismic, and economic problems can be solved. These problems are related to the drilling and casing of superdeep boreholes in high temperature conditions (300-350°C). Other aspects include the development of extensive highly permeable zones that would assure adequate circulation and heating of the heat carrier, and preventing the release of radioactive gases, settling of the surface, and earthquakes. Despite these difficulties, however, the USSR is apparently striving to tap HDR reserves.

The USSR has developed an installation for drilling superdeep boreholes to a depth of 15 km: the Uralmash 15000.* However, a serious hindrance for creating underground boilers at great depths is the very high cost and slow speed of drilling such wells; furthermore, in the case of artificial underground boilers, several wells must be drilled. In order to increase the productivity of drilling operations and reduce their cost, intensive research is being conducted on the development of new drilling techniques (drilling with the aid of explosions, reactive turbo-drilling, use of ultra-sound to break up rock, erosion drilling, etc.). Also being studied is the development of thermally stable drilling muds that can be used in temperatures on the order of 400°C and geophysical equipment for studying wells up to 15 km deep with similar formation temperatures.

The first priority in the USSR's HDR program will likely be the development of artificial boilers at comparatively low depths (up to 3 km) in order to derive low potential hot water for the expansion of the

---

*The deepest borehole on record--the Kol'skaya well--was drilled to a depth of 12 km using this installation. Temperatures of 180°C have been recorded at the well.

mining industry in permafrost regions of Yakutia, Magadan, and Chukotka.
The Avachinskiy volcano region on Kamchatka offers potential for the
development of a geothermic boiler near the volcano's magmatic chamber,
which lies at a depth of approximately 2,000 m. Chamber temperatures
are on the order of 1,000°C, and boiler capacity may represent 250 MW
for a hundred-year period.

In addition to the activity noted above, efficient systems for
collecting the conductive heat in HDR and transforming it into other
types of energy are being sought. The behavior of deep-lying rock in
high temperature and high pressure conditions must be studied, and
effective and economic methods for handling large volumes (cubic kilometers) of crushed rock in HDR formations must be developed. New
materials which facilitate long-term exploitation of geothermal systems
will have to be developed as well. Imposing as these complexities may
seem, HDR should be viewed as a very promising energy resource for the
USSR. Widespread use of HDR energy may be commonplace as early as the
end of this century or the beginning of the next.

3. Summation

Since the First All-Union Conference on Geothermal Research in
1956, the Soviet Union has assembled and actuated an energy production
branch dedicated to the exploration and exploitation of its vast geothermal energy reserves. Despite many significant achievements, however,
this sector is not without its shortcomings. Furthermore, knowledge
gained in the course of almost 30 years has led to a more realistic
assessment of the role geothermal energy might play in the economy of
the USSR. The initial euphoria at the discovery of this "free"
renewable resource has given way to some disappointment and to the

sobering realities of complex obstacles which are as much administrative and economic as they are technical and technological.

In the USSR no separate administrative structure for geothermal energy has been developed. Exploratory efforts for thermal water are still carried out by two ministries: the Ministry of Geology and the Ministry of the Gas Industry, while extraction and sale of thermal water is conducted solely by the latter. The lack of an independent organization hinders the sector's development, resulting in irregularities in planning, distribution of investments, allocation of materials and equipment, etc.; geothermal energy thus must often take the back seat to the needs of natural gas. The creation of an independent economic organization for geothermal energy is apparently viewed as undesirable due to this sector's insignificant contribution to the country's economy. This situation may change when the exploitation of HDR is initiated.

The outlook which has emerged for geothermal energy in the USSR is by no means bleak, however. Grounds for optimism are provided in part by the fact that since the inception of its geothermal program, the Soviet Union has not viewed the resource in narrow terms, i.e. solely as a replacement for fossil-fuel generated electricity. The Soviet experience has borne out the fact that the reserves of heat concentrated in hydrothermal formations are comparatively low and thus will not signify in the fuel balances of industrially developed countries. The USSR regards geothermal water as a multi-purpose resource, and its geothermal program reflects this perception: the focus of Soviet geothermal R&D has included a variety of uses, and will continue to do so in the foreseeable future.

Successes in the use of geothermal water in agriculture, heat and hot water supply, as well as for balneologic purposes testify to the appropriateness of such an approach, and clearly influence the formulation of increasingly ambitious plan targets for geothermal hot water and steam recovery. The great stimulus given to the geothermal program is also visible in the expansion of geologic and exploratory efforts: from 1977 to 1980 they were increased by a factor of 3.5, and they are expected to double from 1980 to 1985.

The future of geothermal energy lies in the recovery of the widely distributed and almost limitless sources of HDR-derived energy. It is not unreasonable to suppose that the development of economically feasible underground geothermal boilers in superheated rock which does not lie deeply below the surface may be accomplished in the next ten years. In this perspective the growth and development of the Soviet geothermal energy sector may be cast in the following three stages: (1) the mid-1960s, in which thermal water in liquid and steam phase began to be utilized; (2) the 1990s, which should see the use of conductive HDR heat in volcanic regions of Kamchatka and permafrost areas of the northeastern USSR via underground thermal boilers lying at depths of up to 3 km; (3) the turn of the century, by which point high-temperature (400-600°C) thermal boilers in HDR at depths of 6-15 km should come into use. Thus, the use of geothermal heat may become an important long-term alternative energy resource in the near future.

ACRONYMS

AR	Apparent resistivity
GGRL	Gamma-gamma ray log
GKZ	State Commission on Mineral Resources
GRL	Gamma-ray log
GWF	Geothermal water field
HDR	Hot dry rock
LLS	Lateral logging sound
NGRL	Neutron-gamma ray log
SP	Self-potential

REFERENCES

1. <u>Proceedings of the United Nations Symposium on the Development and Use of Geothermal Resources</u>. Pisa: 1970, vol. 1-3.

2. <u>Proceedings of the Second United Nations Symposium on the Development and Use of Geothermal Resources</u>. San Francisco: Lawrence Berkeley Laboratory, 1975, vol. 1-3.

3. <u>Trudy pervogo vsesoyuznogo soveshchaniya po geotermicheskim issledovaniyam</u>. Moscow: AN USSR, 1961, vol. 2.

4. <u>Geotermicheskiye issledovaniya i ispol'zovaniye tepla Zemli. Trudy vtorogo soveshchaniya po geotermicheskim issledovaniyam v SSSR</u>. Moscow: Nauka, 1966.

5. Regional'naya geotermiya i rasprastranyeniye termal'nykh vod v SSSR. <u>Trudy vtorogo soveshchaniya po geotermicheskim issledovaniyam v SSSR</u>. Moscow: Nauka, 1967.

6. <u>Izucheniye i ispol'zovaniye glubinnogo tepla Zemli</u>. Moscow: Nauka, 1973.

7. Berman, E. R. <u>Geothermal Energy</u>. Park Ridge, N.J.: Noyes Data Corporation, 1975.

8. Nurshanov, V. A., and B. M. Drozd. Ispol'zovaniye glubinnogo tepla Zemli v narodnom khazyaystve. In: [6].

9. Fomin, V. A., et al. Geothermal Resources of the USSR and Their Geologic-Economic Subdivision. In: [2], vol. 1.

10. Mavritskiy, B. F., and V. G. Khelkvist. Exploration for Thermal Water Fields in the USSR. In: [2], vol. 1.

11. Edwards, L. M., et al. <u>Handbook on Geothermal Energy</u>. Houston: Gulf Publishing Company, 1982.

12. Frolov, N. M., et al. <u>Metodicheskiye ukazaniya po izucheniyu termal'nykh vod v skvazhinakh</u>. Moscow: Nedra, 1964.

13. White, D. E. <u>Geothermal Energy</u>. U.S. Geological Survey Circular 519, 1965.

14. Bogorodistkiy, K. F. <u>Vysoko-termal'nyye vody v SSSR</u>. Moscow: Nauka, 1968.

15. McNitt, J. R. The Geologic Environment of Geothermal Fields as a Guide to Exploration. In: [1], Rapporteur.

16. Mavritskiy, B. F., et al. Resursy termal'nykh vod SSSR. Moscow: Nedra, 1975.

17. Vadetskiy, Yu. V. Bureniye neftyanykh i gazovykh skvazhin. Moscow: Nedra, 1978.

18. Polevoy, S. L. K voprosu rayonirovaniya plastovykh vodonapornykh sistem dlya otsenki perspektiv ispol'zovaniya termal'nykh vod. In: Geologiya, bureniye i razrabotka gazovykh mestorozhdeniy Predkavkaz'ya. Moscow: Nedra, 1967.

19. Tishlyar, I. S., and S. L. Polevoy. Metod ekonomicheskoy otsenki ekspluatatsii i ispol'zovaniya geotermal'nogo mestorozhdeniya. In: [18].

20. Polevoy, S. L., I. P. Kovshov, and I. L. Lumelskiy. Obrabotka rezul'tatov opytnykh vypuskov termal'nykh vod iz skvazhin. In: Razvedka i Okhrana Nedr 11, 1968.

21. Polevoy, S. L., and I. P. Kovshov. Otsenka prognoznykh ekspluatat-sionnykh zapasov termal'nykh vod v usloviyakh vodonapornykh sistem (na primere Predkavkaz'ya). Izvestiya vysshikh uchebnikh zavedeniy. Seriya "Geologiya i Razvedka" 10, 1968.

22. Polevoy, S. L., V. I. Savchenko,, and V. V. Lisovin. Opredeleniye gidrodinamicheskikh parametrov po dannim oprobovaniya skvazhin. Razvedka i Okhrana Nedr 10, 1972.

23. Lumelskiy, I. L., and S. L. Polevoy. Novyye dannyye o zakonomer-nostyakh razmeshchenya i gidrogeologii nekotorikh mestorozhdeniy termal'nikh vod Tsentral'nogo i Vostochnogo Predkavkaz'ya. Trudi po geologii i poleznym iskopayemym Severnogo Kavkaza 13, 1972.

24. Polevoy, S. L. Opredeleniye koeffitsiyenta teploprovodnosti gornikh porod v yestestvennikh usloviyakh. Izvestiya visshikh uchebnikh zavedeniy. Seriya "Geologiya i Razvedka" 10, 1972.

25. Polevoy, S. L., and I. P. Kovshov. Poiskovo-razvedochnyye raboty na termal'nyye vody v usloviyakh plastovikh vodonapornikh sistem (na primere vostochnogo Predkavkaz'ya). In: [6].

26. Polevoy, S. L. Issledovaniye i otsenka mestorozhdeniy termal'nykh vod v usloviyakh plastovykh sistem na primere Vostochnogo Predkavkaz'ya (Avtoreferat kandidatskoy dissertatsii). Novocherkassk: 1969.

27. Polevoy, S. L., I. P. Kovshov, and V. G. Senkevich. Obosnovaniye ploshchadey dlya bureniya geotermal'nykh skvazhin na teritorii Vostochnogo Predkavkaz'ya (Unpublished manuscript). SevKavNIIgaz Library, Stavropol: 1966.

28. Sukharev, G. M. Gidrogeologiya neftyanykh i gazovykh mestorozh-deniy. Moscow: Nedra, 1971.

29. Polyak, B. G. Geotermicheskiye osobennosti oblasti sovremennogo vulkanizma (na primere Kamchatki). Moscow: Nauka, 1966.

30. Itenberg, S. S. Izucheniye nefte-gazonosnykh tolshch promyslovo-geofizicheskimi i geologicheskimi metodami. Moscow: Nedra, 1967.

31. Litvinov, A. A., and A. F. Blinov. Promyslovyye issledovaniya skvazhin. Moscow: Nedra, 1964.

32. Al'tshul', A. D. Gidravlicheskiye poteri na treniye v truboprovodakh. Moscow: Gosenergoizdat, 1963.

33. Bochever, F. M., and N. N. Verigin. Metodicheskoye posobiye po raschetam ekspluatatsionnykh zapasov podzemnykh vod dlya vodosnabzheniya. Moscow: Gosstroyizdat, 1961.

34. Bochever, F. M. Raschety ekspluatatsionnykh zapasov podzemnykh vod. Moscow: Nedra, 1968.

35. Maksimov, V. A. O neustanovivshemsya pritoke uprugoy zhidkosti k skvazhinam v neodnorodnoy srede. Prikladnaya matematika i tekhnicheskaya fizika 3, 1962.

36. Guseyn-Zade, M. A. Osobennosti dvizheniya zhidkosti v neodnorodnom plaste. Moscow: Nedra, 1965.

37. Shchelkachev, V. N. Razrabotka neftevodonosnykh plastov pri uprugom rezhime. Moscow: Gostoptekhizdat, 1959.

38. Bindeman, N. N., and F. M. Bochever. Regional'naya otsenka ekspluatatsionnykh zapasov podzemnykh vod. In: Sovetskaya Geologiya 1, 1964.

39. Krashin, I. I., and D. I. Peresun'ko. Otsenka ekspluatatsionnykh zapasov podzemnykh vod metodom modelirovaniya. Moscow: Nedra, 1976.

40. Healy, I. Pre-Investigation Geological Appraisal of Geothermal Fields. In: [1], vol. 2 (part 1).

41. Gil'denson, M. I. K voprosu ob okhlazhdenii nefti v stvole skvazhine pri yeye dvizhenii ot zaboya k ust'yu. In: Trudy Kuybyshevskogo NII neftyanoy promyshlennosti 11, 1960.

42. Goleva, T.A., Ed. Gidrogeologiya SSSR. Kamchatka, Kuril'skiye i Komandorskiye ostrova. Moscow: Nedra, 1972.

43. Galvina, G. B., and F. A. Makarenko. Geothermal Map of the USSR. In: [2], vol. 2.

44. Lyal'ko, V. I. Balans termal'nykh vod Zakarpat'ya i prognozirovaniye ikh zapasov. In: [6].

45. Beder, B. A. Resursy termal'nykh vod Sredney Azii. Skhema ikh rasprostraneniya i puti ratsional'nogo ispol'zovaniya. In: [6].

46. Grebenshchikova, T. B., G. V. Kulikov, and L. Kh. Safayeva. Gidrotermal'nyye resursy Uzbekistana. In: [6].

47. Mavritskiy, B. F., and G. K. Antonenko. Prognoznyye ekspluatatsionnyye zapasy termal'nykh vod Zapadno-Sibirskogo artezianskogo basseyna i perspektivy ikh osvoyeniya. In: [6].

48. Mizinov, N. V. Tobol'skyye mestorozhdeniya termal'nykhvod i ikh prakticheskoye ispol'zovaniye. In: [4].

49. Lysak, S. V. Geotermicheskiye usloviya i termal'nyye vody yuzhnoy chasti Vostochnoy Sibiri. Moscow: Nauka, 1968.

50. Teplo Zemli i ego izvlecheniye. Kiev: Naukova Dumka, 1974.

51. Naboko, S. I. Pauzhetskiye geyzery. In: Byulleten' vulkanologicheskoy stantsii 22, 1954.

52. Dunichev, V. M. Parogidrotermy Goryachego Plyazha i perspektivy ispol'zovaniya termal'nykh vod Kuril'skikh ostrovov. In: [6].

53. Dvorov, I. M. Geotermal'naya energetika. Moscow: Nauka, 1976.

54. Vakin, Ye. A. Gidrogeologiya sovremennykh vulkanicheskikh struktur i gidrotermal'nyye sistemy yugo-vostoka Kamchatki (Candidate dissertation). Moscow: AN USSR Institute of Geology, 1968.

55. Samoylenko, P. I., V. N. Popov, and G. Z. Avdeyeva. Rezul'taty issledovaniya termal'nykh vod Kamchatki i perspektivy ikh ispol'zovaniya. In: [6].

56. Gavronskiy, A. A. Ekonomicheskiya effektivnost' parovodyanykh skvazhin Pauzhetskogo mestorozhdeniya. In: [4].

57. Kutateladze, S. S. Nauchnyye i prakticheskiye meropriyatiya po razvitiyu geotermal'noy energetiki v Sovetskom Soyuze. In: [4].

58. Kozlov, B. K. Energeticheskoye ispol'zovaniye termal'nykh vod. In: [6].

59. Makarenko, F. A., and V. I. Kononov. Gidrotermal'nyye rayony SSSR i perspektivy ikh osvoyeniya. In: [6].

60. Gadzhiev, A. Skovannaya energiya. Pravda, July 5, 1981.

61. Kozlovskiy, Ye. A. Novyye rubezhi geologov. In: Razvedka i Okhrana Nedr 1, 1981.

62. Dobrynin, P. A., and P. A. Kulik. Dal'neyshee usileniye rabot--glavnaya zadacha gidrogeologov. In: Razvedka i Okhrana Nedr 6, 1981.

63. Natanov, Kh. Kh., and P. A. Kulik. Stabilizatsonnaya obrabotka termal'noy vody v vikhrevykh reaktorakh. In: Geotermicheskiye issledovaniya v Dagestane i voprosy prakticheskogo ispol'zovaniya tepla Zemli. Makhachkala: Dagestan AN SSSR, 1970.

64. Mavritskiy, B. F., B. A. Lokshin, and A. V. Vol'fenfel'd. Progno-
    znyye zapasy termal'nykh vod SSSR, vozmozhnyye ob"yemy vnedreniya
    geotermal'nogo teplosnabzheniya. In: [6].

65. Sivashinskiy, I. I. Ispol'zovaniye termal'nykh vod dlya
    teplosnabzheniya zashchishchennogo grunta. In: [6].

66. Carlson, R. H. Utilizing Nuclear Explosives in the Construction
    of Geothermal Power Plants. In: Proceedings of the Second
    Plowshare Symposium. UCRL-5677, 1959.

67. Kozlovskiy, Ye. A. Vsemirnyy Forum Geologov. In: Nauka i Zhizn'
    10, 1984.

The Delphic Emigre Series

Since 1970, the USSR has allowed the departure of a great number of its citizens. The US public is aware of a large number of Jewish emigres--exceeding 200,000--of whom about half have settled in the United States. There has been, however, an unnoticed exodus of ethnic Germans and Armenians, as well as members of other national groups. These emigres provide an opportunity for enriching our understanding of the Soviet Union, offering us a view of conditions as they really are, not filtered by censorship or embellished to serve political purposes.

Delphic Associates has selected a few emigres with unusual educational and professional backgrounds to prepare scholarly studies on facets of the Soviet Union not well understood by the professional community or academia in the West. To date, this program has resulted in the following monographs:

Fyodor Kushnirsky	Soviet Planning: Evolution in 1965-1980
Victor Yevsikov	Re-Entry Technology and the Soviet Space Program
Sergei Zamascikov	Political Organizations in the Soviet Armed Forces: The Role of the Party and Komsomol
Fyodor Kushnirsky	Price Inflation in the Soviet Machine-Building and Metalworking Sector
Bella Feygin	The Theory and Practice of Price Formation in the USSR
Mark Goldberg	The Development of Combinatorics in the USSR
Avgustin Tuzhilin	The Theory and Application of Wave Propagation and Diffraction in the USSR
Mikhail Turetsky	The Introduction of Missile Systems into the Soviet Navy (1945-1962)
Irina Dunskaya	Security Practices at Soviet Scientific Research Facilities
Esfir Raykher	Natural Gas Transport in the USSR
Sergei Solntsev	The Soviet Aluminum Industry: Methods and Materials
Avgustin Tuzhilin	Systems Analysis and Mathematical Modeling in the USSR
Isaak Adirim	Stagflation in the USSR

Bella Feygin	Economics and Prices in the Soviet Fuel and Energy Industry
Yefim Rivkin	The Metallurgy of Soviet High-Speed Diesels
Samuel Heifets	Stochasticity Research in the USSR: The Novosibirsk Institute of Nuclear Physics
Esfir Raykher	The Economics of the Soviet Gas Industry
Petre Nicolae	CMEA in Theory and Practice
Sergei Polikanov	Nuclear Physics in the Soviet Union: Current Status and Future Prospects
Valentin Litvin	The Soviet Agro-Industrial Complex: Structure and Performance
Vily Khazatsky	Industrial Computer-Based Real-Time Control Systems in the Soviet Union
Vladimir Kresin	Low Temperature Physics Research in the Soviet Union
Savely Polevoy	Geothermal Energy in the USSR: A Survey of Resources, Methodology, Geology, and Use
Eugene Mogendovich	Hydropulse Systems and Technology in the USSR
Lev Chaiko	Helicopter Construction in the USSR
Gabriel Jakobson	Soviet Artificial Intelligence Research: The Tallinn Institute of Cybernetics
Vladimir Shlapentokh	Sociology and Politics: The Soviet Case

www.ingramcontent.com/pod-product-compliance
Lightning Source LLC
Chambersburg PA
CBHW081149180526
45170CB00006B/1992